Why Aren't They Here?

The Question of Life on Other Worlds

Surendra Verma

ICON BOOKS

Published in the UK in 2007 by
Icon Books Ltd, The Old Dairy, Brook Road,
Thriplow, Cambridge SG8 7RG
email: info@iconbooks.co.uk
www.iconbooks.co.uk

Sold in the UK, Europe, South Africa and Asia
by Faber & Faber Ltd, 3 Queen Square,
London WC1N 3AU or their agents

Distributed in the UK, Europe, South Africa and Asia
by TBS Ltd, TBS Distribution Centre, Colchester Road
Frating Green, Colchester CO7 7DW

This edition published in Australia in 2007
by Allen & Unwin Pty Ltd, PO Box 8500,
83 Alexander Street, Crows Nest, NSW 2065

Distributed in Canada by
Penguin Books Canada, 90 Eglinton Avenue East,
Suite 700, Toronto, Ontario M4P 2YE

ISBN-10: 1-84046-806-8
ISBN-13: 978-1840468-06-9

Li   Printed and   )esign

Contents

List of illustrations

Surendra Verma is a science journalist based in Melbourne, Australia. He is the author of *The Mystery of the Tunguska Fireball* (Icon, 2005) and *The Little Book of Scientific Principles, Theories & Things* (New Holland, 2005; Sterling Publishing, 2006). He has also written *The Cause of Mosquitoes' Sorrow* (Icon, 2007), a book about scientific breakthroughs, beginnings and blunders.

The Universe, Life and Fermi's Paradox

The universe. A 13.7-billion-year-old expanse of flat space filled with hundreds of billions of galaxies and vast clouds of glowing gas called nebulae (Latin for 'clouds'). This infinite space has been expanding and the galaxies are moving further apart. The rate of expansion is mind-boggling: imagine a pea growing to the size of the Milky Way in less time than it takes to blink. Only a tiny proportion of the universe's stuff is visible. The rest is known only as dark matter and dark energy. Their true nature is still a mystery. Clues to the destiny of the universe are hidden in dark matter and dark energy.

The Milky Way. A 13.6-billion-year-old spiral disc of about 250 billion stars and huge clouds of gas and dust. This pancake-like disc is more than 100,000 light years wide (a light year is the distance that light travels in one year – about 9,500 billion kilometres). The spiral arms harbour young stars and are therefore brighter than the regions in between. A skinny bar of billions of relatively old, dim red stars, 27,000 light years long, cuts across the heart of the galaxy. There is no dark syrupy chocolate at the centre of this Milky Way bar, but a dark, fiery monster: a super-massive black hole, an infinitely dense point in space-time, sucking in everything around it. Every twinkling star we can see with the naked eye is part of our galaxy.

The solar system. A 4.6-billion-year-old star and its planets in one of the spiral arms of the Milky Way, about 26,000 light years from the galaxy's centre. The Sun is simply a sphere of burning gas, mostly hydrogen. A perfect example of elegance in simplicity.

The eight planets. Outwards from the Sun: Mercury (with a furnace-like temperature of almost 500 degrees Celsius, it's the hellhole of the solar system); Venus (seems like Earth's twin, but it's a parched and scorched world); Earth (liquid water, the elixir of carbon-based life, covers 75 per cent of its surface); Mars (it's no longer a scientific heresy to say: 'Life, not of the little-green-men variety, exists on the red planet today'); Jupiter (one of its 63 known moons, Europa, fascinates astronomers more than the largest planet, as it may have a deep ocean beneath its icy surface); Saturn (rainbows form after it rains on the ringed planet's largest moon, Titan, but its 'water' is liquid methane; geysers of water spew out of reservoirs just metres beneath the icy surface of its tiny moon, Enceladus, hinting at the possibility of life); Uranus (a blue-green wonder with eleven separate rings and 27 known moons); and Neptune (the smallest of the four 'gas giants', so called because they do not have solid surfaces – the other three are Jupiter, Saturn and Uranus).

The dwarf planets. The most famous is Pluto (the 'ninth planet' has now been relegated to this new class of objects); others are Ceres and Eris (formerly called Xena, after the warrior princess of TV fame), and many more icy balls waiting to be classified. And then there are comets and asteroids, debris left over from the birth of the solar system, and (some astronomers speculate) up to a dozen planets, all bigger than Mars.

Planets outside our solar system. So far, more than 200 extra-solar planets have been discovered. Most of them are gas giants, but scientists are enthusiastic about discovering Earth-like worlds with oceans, continents and mountains – and probably some form of life.

The ages-old question 'Are we alone?' has now become tantalising.

In 1950 Enrico Fermi answered the question in his own inimitable way: by asking another question. Fermi, the greatest Italian scientist of modern times, was forced to flee Italy shortly after receiving the 1938 Nobel Prize for physics for his work on nuclear processes. He moved to the United States, where in 1942 he built the world's first nuclear reactor.

Fermi revelled in posing unexpected questions on aspects of the natural world and then figuring out their answers. According to American physicist Philip Morrison: 'Fermi was the first physicist to my knowledge who enjoyed doing physics out loud walking through the hall.' He describes an incident when they were walking through the wooden barrack-like structure of the theoretical physics building at Los Alamos Laboratory, and as they walked, the sounds of their footsteps reflected off the wooden surface and seemed to bounce throughout. Fermi asked Morrison: 'How far do you think our footsteps can be heard in this building?' Fermi then quickly started working out loudly what the yield of sound would be from the impulse, how far that would go, how the wood conduction and air passage would affect the sound, and by the end of the hall he had an answer. 'Sounded very reasonable,' says Morrison. 'And when I tried to recalculate, I got something like the same result – slowly and looking at the numbers over and over again.'

Questions that can be answered quantitatively by rough approximations, inspired guesses and statistical estimates from very little data are known as Fermi questions. Some classic Fermi questions: How many piano tuners are there in the city of Chicago? How many atoms could be reasonably claimed to belong to the jurisdiction of the United States? How far can a crow fly? These questions can be answered by making reasonable assumptions, not necessarily relying upon definite knowledge for an exact answer. (You can find many Fermi-type questions by Googling 'Fermi questions'. In the meantime, how many copies of this book would you need to fill your room?)

At a lunch in the summer of 1950 at Los Alamos, Fermi and fellow nuclear physicists Emil Konopinksi, Edward Teller and Herbert York were talking about space travel. The discussion was probably prompted by a recent cartoon in the *New Yorker* magazine explaining why rubbish bins were disappearing from the streets of New York City. The cartoon showed 'little green men' with antennae carrying rubbish bins towards a flying saucer.

The discussion veered towards the possibility of many civilisations beyond Earth. Fermi surprised everyone by asking the provocative question: 'If they are there, why aren't they here?' This is Fermi's paradox.

There are many explanations: *serious* (evolutionary biologist Ernst Mayr: the evolution path that leads to intelligent life is far more complex than we suppose – we are, if not the first, then among the first intelligent life forms to evolve in the galaxy; astronomer Carl Sagan: daunting distances of interstellar space make space travel impossible …

if we are alone in the universe, it sure seems like an awful waste of space); *bizarre* (astronomer John A. Ball's zoo hypothesis which portrays Earth as a zoo of an intelligent life in the galaxy – they're watching us from a distance); *humorous* (science writer Arthur C. Clarke: 'I'm sure the universe is full of intelligent life – it's just been too intelligent to come here.'); and *optimistic* (astronomer Frank Drake: 'They could show up here tomorrow.').

The question of life on other worlds has two aspects: a) is there any life form, even microscopic, on a planet other than our own; and b) is there intelligent life, able to communicate with us, on a star-system other than our own? Both aspects have profound scientific, philosophical and theological implications.

What follow are glimpses of extraordinary ideas of extraordinary minds about things extraterrestrial – presented in a non-technical narrative accessible to everyone.

The understanding between a non-technical writer and his reader is that he shall talk more or less like a human being and not like an Act of Parliament. I take it that the aim of such books must be to convey exact thought in an inexact language ... he can never succeed without the co-operation of the reader.

Arthur Eddington, British astrophysicist, 1935

CHAPTER I

Histories

Science's poster boy of antiquity

In our times, science's unlikely poster boy is a wild-haired old man with piercing eyes. In times which are tagged in our history books as BC (or BCE – before Common Era – in politically correct language), Einstein's equivalent was Aristotle.

Even half a century after his death, Einstein's face – with its unique mixture of intensity and distraction – is familiar to billions of people around the world. In days when there was no TV, no internet, not even newspapers, Aristotle's face never became ubiquitous. However, his influence was extraordinary: his ideas dominated scientific thought for nearly 2,000 years. A feat yet to be surpassed by Einstein!

But what did he look like? Ancient portrait busts show a stern, sage-like figure: a little bald-headed, a deep brow, small eyes, an aquiline nose and a thick beard. Writing five centuries after Aristotle's death, Diogenes Laertius, a 2nd-century AD Greek biographer, remarked: 'He had a lisping voice ... he wore fashionable clothes, and rings on his fingers, and used to dress his hair carefully.' Very much unlike Einstein.

When Aristotle was born in 384 BC in Stagira, a small town in northern Greece, the Greek world was more than Greece – it extended along the coasts of Turkey, Egypt, Libya, Spain, France and Italy – and was composed of self-governing territories with a single urban centre (the *polis*, the city-state, from which comes 'politics'). The largest *polis* was Athens, with a population of about 300,000.

At the age of seventeen, Aristotle travelled to Athens where he enrolled in the Academy of the renowned philosopher Plato. The Academy was not like a college or university of today: there were no formal courses, no examinations, no degrees. Instead, there were discussions, mostly in the form of a dialogue between two people. In a dialogue, one person asks a question, the other expands upon the idea, presents possible answers, and asks further questions, and so on. This unique method of furthering knowledge was pioneered by Plato's teacher, the philosopher Socrates, who was accused of corrupting the minds of young Athenians and was sentenced to death by drinking poison.

Young Aristotle blossomed in the Academy's stimulating intellectual environment; Plato nicknamed him 'the reader' and 'the mind of the Academy'. A keen reader he was, indeed. Diogenes confirms this: 'He always placed beside his bed a brass dish, and when he lay down to read and rest, he would hold in his hand extended over the dish a brass ball. When he was overcome by sleep, the ball would fall into the dish, and the noise would immediately awaken him.' The use of this Aristotelian contraption is not compulsory while reading this book.

Aristotle stayed at the Academy for about twenty years, first as a student and then as a teacher. In 343 BC, King Philip of Macedonia, who was Aristotle's boyhood friend, asked him to tutor his thirteen-year-old son, Alexander (the future Alexander the Great). Aristotle taught the young prince for three years, and in 334 BC returned to Athens to set up his own academy at the Lyceum, a temple of the sun god Apollo. Since Aristotle had the habit of walking with his students up and down the academy's main avenue while teaching and discussing, the academy became known as the Peripatetic School (from the Greek *peripatētikos*, 'walking up and down').

After Alexander's death in 323 BC in Babylon, anti-Macedonian sentiments in Athens began to rise. Aristotle, who came from the area, was charged with impiety, the lack of reverence for religion. Rather than suffer the fate of Socrates he fled to Chalcis, his mother's home town, saying that he would not allow Athens to sin twice against philosophy. He died the next year, 322 BC, leaving an astonishing legacy for posterity: a corpus that covered most branches of knowledge. Much of this greatest 'encyclopaedia' of the ancient world represented Aristotle's original thoughts and observations.

A certain grandeur

In his biography, Diogenes lists some 550 books by Aristotle. Of these, only about a third has survived. By our standards these are rather slim volumes and would make about 2,500 modern pages. Much of this work deals with logic, ethics, politics, metaphysics (the ideas of existence, truth and knowledge), biology, physics and astronomy.

Aristotle's writings on logic are called *Organon* ('the instrument'). They set out his ideas of science as a logical system in which the truths of nature can be deducted from universal principles. These universal principles were perceived intuitively as correct; but, like axioms of geometry such as 'a straight line can be drawn between two points', were unprovable. For example, from the universal principle that the circle was the perfect shape in nature, Aristotle deduced that the heavenly bodies must move in circles. Such was Aristotle's influence that when in 1609 Kepler calculated that the planets moved in elliptical orbits he had difficulty in convincing himself that this was true.

Plato advocated that since the sphere was the perfect shape for a body, Earth must be a sphere. Aristotle presented several scientific and decisive arguments, including the fact that during an eclipse of the Moon, the edge of Earth's shadow was always circular, and this could not happen unless Earth was spherical. He said that it was at rest at the centre of the universe, with the Moon, the Sun and the planets revolving around it in concentric spheres. These spheres were bounded by a celestial sphere, on the concave side of which were the fixed stars. This sphere revolved on its axis once in 24 hours. The stars were spherical and did not have any independent motion. They were not hot; their light and heat came from friction with air. Meteors and comets were not celestial objects because they did not have circular orbits like planets; they were related to meteorological phenomena such as lightning and rainbows.

He suggested that everything on Earth was made from four elements – earth, water, air and fire. Each of these elements had its own type of movement, in a certain direction: earth went down, fire went up, water went above earth, and air went above water but below fire. As the Moon, the Sun and all the other planets and stars travelled in a circle, different from the straight-line motion of the four elements, he proposed that they were made of a fifth element, ether, whose natural movement was circular. The celestial spheres were moved by a divine force which he called the prime mover (in medieval illustrations of the cosmos, the prime mover was transformed into angels).

In the 5th century BC Leucippus and his pupil Democritus, the Greek philosophers famous for their atomic theory – that matter is made of empty space and an infinite number of tiny particles called *atomos* or atoms which are always in motion – also suggested that atoms produced innumerable worlds of different sizes. 'In some worlds there is no sun or moon, in others they are larger than in ours, and others have more than one,' Democritus said. His comment that 'some of the worlds have no animals and plants and no water' is the first-ever scientific allusion to life beyond Earth.

A century later, Epicurus advanced the atomic theory of Democritus and combined it with the pleasure ethics of Aristippus, a pupil of Socrates who believed that sensory enjoyment was the most important thing in life. His followers were called Epicureans and they lived in a garden. The story goes that an inscription on the gate to the garden said: 'Stranger, here you will do well to tarry; here our highest good is pleasure.'

Diogenes records a letter of Epicurus to Greek historian Herodotus in which he wrote: 'There are infinite worlds both like and unlike this world of ours. For the atoms being infinite in number, as was already proved, are borne on far out into space. For those atoms which are of such nature that a world could be created by them or made by them, have not been used up either on one world or a limited number of worlds ... So that there nowhere exists an obstacle to the infinite number of worlds.' Later in the letter he suggests that these worlds have 'living creatures and plants and other things we see in this world'.

Metrodorus of Chios, a pupil of Epicurus, supported his

teacher: 'To consider Earth the only populated world in infinite space is as absurd as to assert that in an entire field sown with millet, only one grain will grow.' His claim that the worlds are infinite in number followed from the atoms being infinite.

In the 1st century BC Lucretius, a Roman poet who was a contemporary of Julius Caesar, wrote an epic poem, *De rerum natura* (On the Nature of the Universe), which kept alive Epicurus' atomism. The main theme of the poem is that behind all natural phenomena lie eternal, unchanging atoms that can arrange and rearrange themselves into different forms. He explains how atoms form solids, liquids and vapours. Solid substances, for example, are 'held together linked and interwoven as though by rings and hooks'.

An interesting aspect of such combinations of atoms is that, as there is an abundance of atoms available, these atoms are capable of forming other worlds elsewhere in the universe, with races of different humans and animals:

> Now if there is so vast a store of seeds
> That the whole lifetime of all conscious beings
> Would fail to count them, and if likewise Nature
> Abides the same, and so has power to throw
> The seeds of things together everywhere,
> In the same manner as they were thrown together
> Into our world, then you must needs admit
> That in other regions there are other earths,
> And diverse stocks of men and kinds of beasts.

Lucretius, *De rerum natura*, trans. R.C. Trevelyan, Cambridge University Press, 1937

Aristotle's elaborate ideas redefined the cosmos, but it remained finite and changeless: he did not entertain the idea of never-ending sequences of celestial spheres. Nor did he entertain the idea of the existence of innumerable worlds advanced by Democritus and Epicurus. In his cosmos, the natural movement of the element earth was towards the centre of the Earth, which was also the centre of the world; but the element fire moved away from the centre of the Earth and towards the outer circumference. The elements air and water assumed their natural places between the centre of the Earth and its outer circumference. If there were more than one world, the elements would have more than one natural place towards which to move. Many worlds suggested the idea of more than one centre and one circumference. To him it was a logical contradiction. 'There cannot be several worlds,' he declared. That ended the debate for centuries.

Aristotle is still admired as a great philosopher, but in matters of science the poster boy of antiquity was wrong most of the time. Unlike Einstein ('It's the theory that decides what we can observe'), he observed without forming detailed theories. His power of observation was distorted by preconceived notions. Some say that his ideas stalled the progress of science until they were challenged by Bacon, and proved wrong by Copernicus, Kepler, Galileo, Newton and others.

Francis Bacon, an English philosopher, advanced a new method of inquiry, completely different from the philosophical methods of Aristotle, in his book *Novum Organum* ('the new instrument'), which has influenced every scientist since its publication in 1620. 'There are two methods of investigation, through argument and through experiment,' he wrote. 'Argument does not suffice, but experiment does.'

He rejected Aristotle's deductive approach to reasoning, and suggested an *inductive* approach. The most important aspect of this method was the idea of drawing up tentative hypotheses from available data and then verifying them by further investigations.

Although Bacon dethroned Aristotle, he did not challenge Aristotle's belief in one world. However, he invoked a different argument: the impossibility of a void between different worlds. 'If there were another universe, it would be of spherical figure,' he said. 'Therefore they must touch; but they cannot touch each other except in one point ... Hence elsewhere than in that point there will be a vacant space between them.'

Bacon discovered the most important tool of science – the scientific method – but he did not make any significant scientific discovery. 'I shall content myself to awake better spirits like a bell-ringer, which is first up to call others to church,' he once wrote to a friend. Bacon's bell is still ringing, yet it can never drown out the contribution of one of the greatest geniuses in the progress of science.

Aristotle said: 'The search for truth is in one way hard and in another way easy, for it is evident that no one can master it fully or miss it wholly. Each adds a little to our knowledge of nature, and from all the facts assembled there arises a certain grandeur.'

No other ancient institution has contributed more to that certain grandeur than the Great Library and Museum at Alexandria.

From poster boy to poster girl

In 331 BC, Alexander stopped during one of his journeys at the western end of the Nile delta, a few kilometres inland from the Mediterranean where the sea turned into the shoreline to form a natural harbour, and decided to found a city there. He took a personal interest in planning it, and marked out various sites, including the Museum (literally a shrine to the Muses), with chalk. Legend has it that when he ran out of chalk, he used barley from the soldiers' mess as a substitute. But it was soon gobbled up by seagulls. Alexander considered it a bad omen but his personal seer, Aristander, assured him that the birds' coming was most auspicious. It was a sign that the new city would be great and prosperous. Aristander's prophesy was fulfilled: Alexandria became the cultural and commercial hub of the ancient world.

After Alexander's death in 323 BC his empire broke into three regions which were ruled by his generals. One of them, Ptolemy Soter, declared himself king of Egypt in 304 BC. The Ptolemaic dynasty ruled Egypt, with its capital at Alexandria, until the death of Cleopatra in 30 BC.

Following Alexander's wishes, Ptolemy Soter founded the Museum. As he had considerable intellectual interest, he also founded a library with the Museum. Demetrius, an Aristotelian philosopher, was put in charge of collecting books for the new library. It was almost certainly modelled on Aristotle's personal library. Not for nothing was Aristotle called 'the reader'. His great collection of books at the Lyceum was arguably the first university library in history. Strabo, a 1st-century AD Greek historian, claims that Aristotle was 'the first man whom I know to have collected books and to have taught the kings of Egypt how to arrange a library'.

Over the centuries the Library amassed more than 400,000 scrolls, but none has survived. This great loss has been attributed, in turn, to the Romans under Julius Caesar in 47 BC (40,000 volumes were burned in a fire lit by soldiers to clear the wharves to block the fleets of Cleopatra's brother),

Christian zealots in AD 391 (an unknown number of works were lost during their riot), and the Arab armies of the Caliph of Baghdad in AD 641 (they not only razed the Library buildings, but burned the books to heat the public baths).

The Library also supported a *synodos* (community) of scholars who made it the chief centre of science and scholarship in the ancient world. Many of the scholars enhanced Aristotle's image of the cosmos.

Aristarchus was a contemporary of Archimedes (who gave us the first 'Eureka!' moment in science), and both were associated with the Library. Aristarchus is credited by Archimedes as having taught that the Earth was not at the centre of the universe, but that it moved around the Sun. His idea of a moving Earth looked utterly strange during those ancient days and it was, of course, rejected by his contemporaries.

Eratosthenes, who was appointed head of the Library in 236 BC, calculated the size of Earth. His value for the diameter, in modern units, was 12,633 kilometres, which is only 101 kilometres short of the true average diameter – a remarkably accurate measurement. He was a versatile scholar: an astronomer, mathematician, geographer, historian, literary critic and poet. He was nicknamed 'Beta' (the second letter of the Greek alphabet) because he was considered the second best at everything.

Hipparchus, in the 2nd century BC, applied a yardstick to the heavens. He used sundials, water clocks and a circular instrument divided into degrees and fitted with a simple sighting device for a rigorous study of the sky. He drew up tables of the motions of the Sun, the Moon and the planets, and from these tables he predicted eclipses for long periods ahead.

'Earth does not rotate; otherwise objects will fling off its surface like mud from a spinning wheel. It remains at the centre of things because this is its natural place – it has no tendency to go either one way or the other. Around it and in successively larger spheres revolve the moon, Mercury, Venus, the sun, Mars, Jupiter and Saturn, all of them deriving their motion from the immense and outermost spheres of fixed stars.' So proclaimed Claudius Ptolemy (not related to the Ptolemys) in c. AD 150 in his book *Almagest*, in which he synthesised the work of his predecessors. A major part of *Almagest* ('the greatest' in Arabic) deals with the mathematics of planetary motion. Ptolemy explained the wanderings of the planets by a complicated system of cycles and epicycles, which harassed astronomers for centuries. 'If the Lord Almighty had consulted me before embarking upon the Creation, I should have recommended something simpler,' commented Alfonso the Wise, the 13th-century Spanish king of Castile, a great patron of astronomy. However, Ptolemy's erroneous theory dominated astronomy for fourteen centuries until it was challenged by Copernicus and demolished by Kepler.

In AD 400 Hypatia, daughter of Theon, a mathematician and astronomer at the Library, became head of the neo-Platonist school of philosophy. She was also an outstanding mathematician and inventor. Only the titles of her mathematical works have survived, but sources describe her as a mathematician who surpassed her father's talents. She invented, among other things, a plane astrolabe to measure the position of stars, planets and the Sun.

Hypatia was a pagan in an increasingly Christian city. One dark night in 415, on the way home from the Library, she was dragged off her chariot by a mob of extremist Christians, stripped stark naked, hacked to death, and her remains burned.

In the 20th century, the life and death of beautiful and brilliant Hypatia was romanticised by feminists and she became the poster girl of modern science.

With Hypatia's death ended the golden age of Greek science. Europe had entered the Dark Ages. Earth – the only world – stood at the centre of a bounded and finite universe. The Christian Church wholeheartedly accepted this Aristotelian view. The first challenge to it came in the 14th century.

Ockham's pen

William of Ockham (also spelled Occam), a philosopher and theologian, came from Ockham, a village in Surrey, near London. In his youth he joined the Franciscan order and studied at Oxford, where he lectured from 1315 to 1319. At Oxford, which was then a great Franciscan centre of learning, William became the leader of a school of philosophy called nominalism.

Aristotle's main argument for a single world was based on the causes of the motions of the four elements. When moved outside its natural place, each element returned to that unique natural place. William, however, suggested that the elements would not necessarily return to their *unique* natural places, but to a natural place dependent on their situation. He rejected Aristotle's view of a finite universe, and said that there could be no assurance that the world was finite, or that it had a governing unity, or that it was eternal, or that there were not several worlds. He said that the elements in each world would return to their natural place within their own world.

William is now most remembered for his rule, known as Ockham's Razor, which is of vital importance in the philosophy of science. The rule – *a plurality (of reasons) should not be posited without necessity, or it is vain to do with more what can be done with less* – implies that the number of causes or explanations needed to account for the behaviour of a phenomenon should be kept to a minimum. It is a guiding principle in developing scientific ideas, and it insists that you should prefer the simplest explanation to fit the facts. The rule has been interpreted in modern times to mean that when you have two competing theories that make exactly the same predictions, the one that is simpler is the better. In other words, the explanation requiring the fewest assumptions is most likely to be correct. Advice to computer programmers to keep their programs simple – *keep it simple, stupid* (KISS) – is in a similar vein. But we must also heed Einstein: 'Everything should be made as simple as possible, but not simpler.'

William's statements in his philosophical and theological writings, including the ideas on other worlds, aroused such opposition that he was refused his Master of Theology degree at Oxford and was ordered to appear before the papal court on charges of heresy. He fled to Germany and, according to a story, probably apocryphal, asked Emperor Louis IV for protection with the plea: 'Protect me with your sword, O Emperor, and I shall protect you with my pen.' He remained in Germany for the rest of his life.

Two centuries later, a pen mightier than that of William redrew Aristotle's cosmos and demoted Earth from its pride of place.

Innumerable worlds

'The sun is at the centre of the solar system, fixed and immobile, and planets orbit around it in perfect circles in the following order: Mercury, Venus, Earth with its moon, Mars,

Jupiter and Saturn,' declared Nicolaus Copernicus, a Polish astronomer and cleric. The Copernican system defied the dogma that Earth stood still at the centre of the universe, and set forth a new theory of a Sun-centred universe.

Not only did Copernicus place the Sun at the centre of the solar system, but he also gave detailed accounts of the motions of Earth, the Moon and the planets that were known at the time. He said that Earth also revolves on its own axis, which accounts for days and nights.

Copernicus had found the truth, but to convince the world was an onerous task. For more than three years he taught his new theory to his students at the University of Rome, but he could not stagger their belief in an Earth-centred universe.

He did not publish his findings because they were thought to contravene the teachings of the Church. Religious leaders of his time were against him. Martin Luther (founder of the Lutheran Church in Germany) denounced him as 'a new astrologer ... the fool' who wanted 'to overturn the entire science of astronomy'. John Calvin quoted Psalm 93 against him: 'Surely the world is established, so that it cannot be moved.'

Copernicus passed his life in agony. His efforts to convince the Church of the truth were in vain. His book *De revolutionibus orbium coelestium* (On the Revolutions of the Heavenly Spheres) was published in Germany at the very end of his life. Andreas Osiander, a Lutheran theologian and one of Copernicus' students, who was supervising the printing process, was so frightened of the revolutionary ideas contained in the book that he wrote a preface claiming that it was not a scientific treatise but a 'playful fancy'. The first copy of the book was sent to Copernicus in Poland. It arrived but a few hours before his death on 23

May 1543. Thus the greatest astronomer of his time died without seeing his book in print – the book which ranks with Newton's *Principia* and Darwin's *Origin of Species* as a product of scientific genius.

Copernicus' book was in defiance of the Church's teachings, but it was too late for the Church to do anything. A few decades later Giordano Bruno, an astronomer and Dominican monk, was burned at the stake for his belief in Copernicus' heretical views.

9 February 1600. It is freezing cold in the vast and ornate Hall of Inquisition in Rome. Fifteen illustrious cardinals of the Holy Office are seated on high-backed plush chairs forming an arc around the accused – a 51-year-old, small, thin man with black hair and dark brown eyes, kneeling silently. The Grand Inquisitor, Cardinal Severina, reads the charges. The eight counts of heresy include belief in the movement of the Earth and in an infinite universe filled with innumerable worlds. Severina asks Bruno to recant his belief and pray for mercy to God. Bruno remains silent. Severina excommunicates the heretic and sentences him to die 'without shedding of blood' (in other words, to be burnt alive at the stake). A defiant Bruno lifts his head and declares: 'Perhaps you, my judges, pronounce this sentence against me with greater fear than I receive it. The time will come when all will see as I see.'

In the early hours of 19 February, Bruno was dragged from his dark, dreary and damp prison cell and asked to put on the sulphur-coloured garb of heresy, covered with pictures of devils and crimson flames and crosses. He was then led in chains through a howling, fanatical crowd to the site of the execution, a public square called the Campo di Fiore (Field of Flowers). The executioner tied him to the stake, piled bundles of sticks up to the chin and placed a torch between his feet. As the flames blazed around him, a priest pushed forward and pressed a crucifix into his hands, but Bruno turned his head away. Within seconds flames seared him, smoke and fire surrounded him. When the fire subsided, his remains were powdered and blown in the wind so that no relic of the heretic would survive. Three centuries later, a statue in honour of Bruno was erected at the exact site where he was burned. The statue, within walking distance of the Vatican, is now surrounded by a colourful and busy market.

Bruno's most important work on the Copernican system is *La cena de le ceneri* (The Ash Wednesday Supper), published in 1584, in which he describes Copernicus as 'a grave spirit, meditative, penetrating and mature; a man who did not surrender himself to any past astronomers.' Bruno also did not surrender himself completely to Copernicus. In *De l'infinito universo e mondi* (On the Infinite Universe and the Worlds), published in the same year, he diverges from Copernicus. This book, which started his fatal dispute with the Church, is in the form of a dialogue between two philosophers named Burchio and Fracastorio. When Burchio asks, 'Then the other worlds are inhabited like our own?', Fracastorio replies: 'If not exactly as our own, and if not more nobly, at least no less inhabited and no less nobly. For it is impossible that a rational being fairly vigilant, can imagine that these innumerable worlds, manifest as like to our own or yet more magnificent, should be destitute of similar or even superior inhabitants.'

Copernicus' ideas are often referred to as the 'Copernican Revolution', but Milič Čapek, an American philosopher, calls this expression an historical inaccuracy because 'it gives Copernicus the credit which really belongs to Giordano Bruno, the first to depart wholeheartedly and consistently from Aristotelian cosmology'. In *E mondi*, when Burchio asks, 'You maintain that Plato is an ignorant fellow, Aristotle an ass and their followers insensitive, stupid and fanatical?', Fracastorio replies: 'As I told you from the first, I regard them as earth's heroes. But I do not wish to believe them without cause, not to accept those propositions whose antitheses are so compellingly true.' 'Who then shall be judge?' asks Burchio. 'Every well-regulated mind and alert judgement', Fracastorio replies.

Descartes' vortices

Fracastorio would have agreed that René Descartes, the 17th-century French philosopher and mathematician who invented analytical geometry ('the method of giving algebraic equations to curves', as Voltaire described it), was indeed a 'well-regulated mind'. Unlike Bruno, he did not criticise Aristotelian physics, but created an entirely new physics to explain the motions of heavenly bodies. The laws of nature were the laws of mechanics, he asserted, and everything in nature could ultimately be reduced to the rearrangement of particles moving according to these laws. Space by itself being nothing, it has no extension – no length, breadth or height. Only matter has the property of extension, and space cannot exist where there is no matter. Matter exists everywhere, and a vacuum exists nowhere.

The most popular aspect of Cartesian physics was the vortex theory. It provided a simple, mechanistic explanation of the universe which was understandable to everyone. The motions of the heavenly bodies, the theory proposed, consist of whirlpools or vortices of a subtle matter – 'ether'. Each vortex has a star surrounded by a vast space. Every star is a sun and every sun has its own family of planets. The universe is infinite and consists of a vast number of vortices, all limiting and circumscribing each other. In the solar system, the planets are carried about in the Sun's vortex and the Moon is carried around Earth in the same way. Descartes was aware of Kepler's laws that the planets move in elliptical orbits; however, his vortices lead to circular orbits.

Descartes was 37 when he presented his vortex theory in an imposing treatise, *Le monde, ou traité de la lumière* (The World, or a Treatise on Light). When he was giving finishing touches to his book, he heard the stunning news that on 22 June 1633 Galileo had been forced by the Inquisition to abjure as a heresy the Copernican doctrine ('I Galileo Galilei, son of Vincenzio Galilei, of Florence ... will not hold, defend or teach the said false doctrine in any manner ... I give assurance that I believe, and always will believe what the Church recognises and teaches as true.'), and had been punished by confinement to his house.

The image of a frail old man of 70 kneeling before the awful tribunal and recanting his belief crushed Descartes. 'He was not only afraid – as any sane man might well have been; he was deeply hurt,' writes E.T. Bell in his authoritative biographies of mathematicians, *Men of Mathematics* (1937). 'He was as convinced of the truth of the Copernican system as he was of his own existence. But he was also convinced of the infallibility of the Pope.' As he had no desire to become a martyr, he decided not to publish *Le monde*: 'Although I consider all my conclusions based on very certain and clear demonstrations, I would not for all the world sustain them against the authority of the Church.'

His masterpiece, *Discours de la méthode* (A Discourse on the Method), which gave analytical geometry to the world, was published in 1637. In 1644 he published his most comprehensive work on physics, *Principia philosophiae*

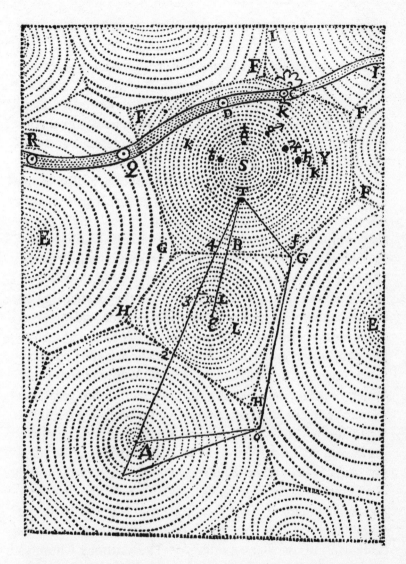

Figure 1. Descartes' system of vortices. Each vortex denotes a star surrounded by a vast space. In the case of the solar system, the vortex carries the planets around the Sun (S). The irregular path across the top of the diagram shows the motion of a comet, which passes through many vortices. (From Descartes' Le monde, ou traité de la lumière)

(Principles of Philosophy), which was an extension of *Le monde*. The Church ignored its publication, but placed it on its Index of prohibited books. Yet it was widely read and caused a stir in the world of science.

In this book Descartes presented the first complete physical system since that taught by Aristotle. The vortex theory caught the imagination of contemporary scientists and was taught widely in universities in Europe, but only until 1687 when Newton presented a revolutionary idea in his magnum opus, *Philosophiae naturalis principia mathematica* (Mathematical Principles of Natural Philosophy), that a single universal force, gravitation, keeps heavenly bodies in their orbits. The vortex theory was now a lost cause. Today it is a mere relic in the museum of dead theories.

The vortex theory obviously implied the existence of an infinite number of planets, yet Descartes remained silent on the issue. When his friend and patron, Christina, Queen of Sweden, was disturbed by the implication that a Cartesian must 'probably hold that all these stars are inhabited, or, still better, that they have earths around them, full of creatures more intelligent and better than he,' he replied guardedly: 'I always leave these questions, once posed, suspended, preferring not to deny and not to affirm anything.' Perhaps the trials of Bruno and Galileo were still fresh in the mind of the man who also gave us the single best-known philosophical statement, *Cogito ergo sum* – 'I think, therefore I am'. Nevertheless, for his followers the trials were as distant as his vortices in the sky. The Catholic Church's authority was waning and the Cartesians were not afraid to assert explicitly the existence of an infinite number of planets, inhabited or otherwise. One of them was Bernard le Bovier de Fontenelle.

The philosopher and the countess

Fontenelle was a poet, playwright, essayist and natural philosopher. He was also an ardent Cartesian. As a writer and as the permanent secretary of the French Academy of Sciences – although he made no scientific discoveries, he was appointed to this prestigious post in 1697 – his influence in furthering the cause of vortex theory was great.

He was a charming and witty man. In his nineties, when he met a young woman, he exclaimed: 'Ah! If I were only eighty now!' He continued to write almost until the day he died – one month short of 100 years – on 9 January 1757. He died without suffering, telling his doctor at his deathbed that he felt nothing apart from 'a difficulty in continuing to exist' ('une difficulté d'être').

Fontenelle's charm and wit – and his admiration for the Copernican system and Descartes' vortices – is most abundant in a lively little book, *Entretiens sur la pluralité des mondes* (Conversations on the Plurality of Worlds). Published in 1686, when the author was only 29, the book became an instant bestseller and is still on sale today. Three different English translations appeared within two years (five more eventually followed, the most recent in 1990). A charming translation is by English poet John Glanvill, *A Plurality of Worlds*, published in 1688 (the latest reprinting was in 1929, a beautifully crafted limited edition by the Nonesuch Press in London; the quotes on these pages are from this edition).

The book was more fiction than physics, but the ideas of other worlds that it displayed were daring, disputable, even disapproved of by the Catholic Church. It was duly placed on the Index of prohibited books a year after publication, removed in 1825 but reinstated in 1900. The Church's

WHY AREN'T THEY HERE?

disapproval could not dampen the soaring popularity of the book, especially among women. (I have placed a woman in these Conversations, Fontenelle writes in the preface, to encourage women through the example of one, 'who without any supernatural parts, or tincture of Learning', never fails to understand what is said to her.)

Fontenelle continued to revise this beloved book, producing 32 editions, the last in 1742, keeping it up to date with the latest scientific discoveries (adjusting the size of Venus or the distance of Saturn, for example, with the most recent astronomical observations).

Written in a playful, whimsical style, *Entretiens* is a delightful, flirtatious conversation between a learned philosopher and a sharp-witted countess that takes place on five consecutive moonlit evenings. The philosopher is convinced that the universe is teeming with intelligent life everywhere. The countess initially scoffs at this bizarre idea, but the philosopher gradually wins her over.

In the first conversation, *Entretiens* introduces the Copernican system: 'Copernicus confounding everything, tearing in pieces the beloved Circles of Antiquity, and shattering their Crystal Heavens like so many Glass Windows: seiz'd with the noble Rage of Astronomy, he snatcheth up the Earth from the Centre of the Universe, sends her packing, and placeth the Sun in the Centre to which it more justly belong ...

'Fairly and softly, *saith the Countess*, I fancy you your self are seiz'd with the Noble Fury of Astronomy; a little less Rupture, and I shall understand you the better.'

The second and third conversations are about the Moon, the planets and their inhabitants: 'Fear it not Madam, *said I*, do you think we are the only Fools of the Universe? Is it not consistent with Ignorance to spread it self every where? ...

Why should Nature be so partial, as to except only the Earth? But let who will say the contrary, I must believe the Planets are peopled as well as the Earth.'

The fourth conversation introduces the complex motion of planets in terms of Descartes' vortices: 'Must my Head, *says she, smiling*, turn round to comprehend 'em, or must I become a perfect Fool to understand the misteries of Philosophy.'

In the last conversation, the philosopher takes the final leap and discusses the fixed stars as suns, around which inhabited planets revolve: 'I perceive, *says the Countess*, where you would carry me; you are going to tell me if the fix'd Stars are so many Suns, and our Sun the centre of a Vortex that turns round him, why may not every fix'd Star be the centre of a Vortex that turns round the fix'd Star? Our Sun enlightens the Planets; why may not every fix'd Star have Planets to which they give light?

'You have said it, *I reply'd*, and I will not contradict you.

'You have made the Universe so large, *says she*, that I know not where I am, or what will become of me; what is it all to be divided into heaps confusedly, one among another? Is every Star the centre of a Vortex, as big as ours? Is that vast space which comprehends our Sun and Planets, but an inconsiderable part of the Universe? and are there as many such spaces, as there are fix'd Stars? I protest it is dreadful.

'Dreadful, Madam, *said I*; I think it very pleasant, when the Heavens were a little blue Arch, stuck with Stars, methought the Universe was too strait and close, I was almost stifled for want of Air; but now it is enlarg'd in heigth and breadth, and a thousand & a thousand Vortex's taken in; I begin to breath with more freedom, and think the Universe to be incomparably more magnificent that it was before.

'Well, *says the Countess*, I have now in my Head, the System of the Universe: How learned am I become?'

Entretiens is more than an immortal classic; it is the first popular-science book ever published, a book that not only became popular but also succeeded in making its scientific message overwhelmingly popular.

'A pleasing hypothesis'

By the beginning of the 19th century the idea of life on other worlds had entered scientists' consciousness. 'Men of science readily accepted a pleasing hypothesis which was not positively contradicted by evidence from their special studies,' writes Isaac Todhunter in his book, *William Whewell* (1876). He cites 'almost a solitary expression of a contrary opinion', a flippant remark in *Table Talk* (1836), the conversations of poet Samuel Taylor Coleridge published after his death in 1834: 'I never could feel any force in the arguments for a plurality of worlds, in the common acceptance of that term. A lady once asked me – "What then could be the intention in creating so many great bodies, so apparently useless to us?" I said – I did not know, except to make dirt cheap.'

The name of English polymath William Whewell looms large in 19th-century science (he coined the word 'scientist' to replace the terms then in use: 'natural philosopher' or 'man of science'). In 1853 he published a book, *Of the Plurality of Worlds: An Essay*, in which he contradicted the 'pleasing hypothesis' that extraterrestrial life was rife in the universe. He presented scientific evidence to show the uniqueness of our planet and the improbability of intelligent life elsewhere. David Brewster, the Scottish physicist, chiefly remembered today for the invention of the kaleidoscope, accused him of depopulating the heavens. By 1859 five editions (each larger than the previous) had been published, and their reviews filled the periodicals. The debate that the *Essay* started lasted for more than a decade.

In fact, that debate has never ended. It will not end until the learned countess has made contact with the philosopher's other worlds.

CHAPTER 2

Fictions

Kepler's dream

Whatever is born from the soil or walks on the soil is of prodigious size. Growth is very quick; everything is short-lived, although it grows to such enormous bodily bulk. They have no settled dwellings, no fixed habitation; they wander in hordes over the whole globe in the space of one of their days, some on foot, whereby they far outstrip our camels, some by means of wings, some in boats pursue the fleeing waters ... Most creatures can dive ... they live deep down under the water.

Who are they? Klingons from planet Kronos, or Wookiees from planet Kashyyyk, or Daleks from planet Skaro ready to 'EX-TER-MIN-ATE!' earthlings? No, they are giant yet harmless Privolvans from the Moon. They are the first 'aliens' to inhabit the first 'other world' created in a science fiction story, a story dreamed more than 400 years ago by a university student.

Kepler was a 22-year-old student at the University of Tübingen in Germany when, in 1593, he dreamed of writing a book: not an astronomical treatise but a science fiction. The idea occurred to him when he decided to devote his graduate thesis to the question: How would the heavenly phenomena appear to an observer on the Moon?

This question, observes Kepler scholar Max Caspar, was inspired by his enthusiasm for the new astronomy espoused by Copernicus, the man whose prophet he was to become. Caspar notes that the idea of the book 'contains the first germ of a work which we shall come to know as the last of the books he published'.

Twelve years passed before Kepler wrote the first draft, and another twelve before he found time to look at it again. During the last decade of his life, from 1621 to 1630, he began to add copious notes to the slender volume (about twenty pages of a modern book). These notes – 223 in all – are four times as long as the text. The book, written in Latin, was published in 1634, four years after Kepler's sudden death.

Somnium sive Astronomia Lunaris (Dream or Astronomy of the Moon) describes a young boy's trip from Earth (named Volva in the book) to the Moon (Lavinia). The boy, Duracotus, lives with his mother Fiolxhilda on an island. Fiolxhilda is a witch who summons up a demon to propel him to the Moon. Kepler is somewhat vague about the method of propulsion, but he is specific about the distance of the Moon – '50,000 German miles' – which is the correct distance if we use the value of a German mile provided in one of his footnotes.

Unlike the ancient Greeks, Kepler knew that the Earth's atmosphere does not extend as far as the Moon. To enable Duracotus to pass through the thinning atmosphere, he is put to sleep with the aid of opiates and his nostrils stopped with moistened sponges. So that the boy's body is not torn apart by the great acceleration needed to escape Earth, he sits with his arms and legs curled inwards. As he is pulled by the demon towards the Moon, he reaches a point 'where the Moon's magnetic force balances that of the Earth'. When the demon releases him, he falls to the Moon unaided. Here Kepler is groping for gravity, well before Newton saw the proverbial apple fall from a tree and hit upon the idea of it.

'Like spiders they will stretch out and contract, and propel themselves forward by their own force – for, as the magnetic forces of the Earth and the Moon both attract the body and hold it suspended, the effect is as if neither of them was attracting it – so that in the end its mass will by itself turn towards the Moon.' That's how he describes the effect of space travel on his hero's body.

'He not only takes it for granted, but, with truly astonishing insights, postulates the existence of "zones of zero gravity" – that nightmare of science fiction,' marvels Arthur Koestler in his magnificent biography of Kepler in *Sleepwalkers* (1959).

When his hero's journey is completed, Kepler proceeds to describe in detail the astronomy, landscape, climate and plant and animal life of both hemispheres of the Moon, called Sublova and Prilova. The story ends when Duracotus is awakened from his dream by a cloudburst.

Somnium skilfully combines a fictional narrative with a treatise on lunar astronomy. It's 'hard' science fiction, the first of its kind. It must be appreciated, as American historian Lewis Mumford says, 'for the audacity of the concept' as well as for its intrinsic merit as a pioneering work of science fiction.

A story of proportions

Kepler went to the Moon to see aliens, but Voltaire brought them to Earth in *Micromégas* ('littlebig' in Greek), a whimsical tale of about 6,000 words. His eponymous hero Micromégas, a 36,000-metre tall, handsome giant with a 15,000-metre waist and a 1,700-metre nose, is the first science fiction alien to visit our puny planet.

Micromégas has been banished from the star Sirius for 800 years for making 'suspicious, offensive, rash and heretical statements' in a book about little insects less than 30 metres in diameter. He decides to travel 'from planet to planet, with a view to improving his mind and soul'. He has such 'a marvellous acquaintance with the laws of gravitation, and all of the forces of attraction and repulsion' that he travels through the Milky Way 'sometimes by a means of sunbeam, and sometimes with the help of a comet'. He arrives at the planet Saturn and befriends a native, who is only 1,500 metres tall, a mere dwarf beside Micromégas. Together they travel throughout the solar system, and when they reach Earth the Saturnian complains that 'this globe is so ill-constructed, so irregular, and so ridiculously shaped ... I cannot suppose any sensible people should wish to occupy such a dwelling'.

Like *Candide*, Voltaire's most popular work, *Micromégas* is a satire. Published in 1752, it makes satirical observations on the pretensions of humans. The visitors to Earth marvel at its smallness and its 'microscopic' inhabitants. This was a revolutionary thought in Voltaire's time, when Earth was believed to be the centre of the universe and humans its perfect creatures.

The Sirian, who has a thousand senses as opposed to the poor Saturnian's 72, notices a shipload of philosophers who are returning from a voyage to the polar circle. Micromégas picks up the ship and places it in the hollow of his hand. He then examines 'the mere mites' with a microscope he has brought with him. The two travellers are amazed to find that the 'invisible insects' can speak and even have the capacity to reason. Once they get over their amazement, they start lengthy conversations with the philosophers.

'O intelligent atoms ... you must doubtless taste joys of perfect purity on your globe; for, being encumbered with so little matter, and seeming to be all spirit, you must pass your lives in love and meditation – the true life of spiritual beings.' The philosophers all shake their heads and one, more frank than the others, replies: 'We have more matter than we need, the cause of much evil, if evil proceeds from matter; and we have too much mind, if evil proceeds from mind. For instance, at this very moment there are 100,000 fools of our species who wear hats, slaying 100,000 fellow creatures who wear turbans, or being massacred by them.'

Micromégas shudders and asks the philosopher the cause of such horrible quarrels. 'The dispute concerns a lump of clay,' replies the philosopher, 'no bigger than your heel.'

We run the risk of being called bore if we tell you more, as Voltaire (pseudonym for François Marie Arouet) once said: 'The secret of being a bore is to tell everything.' ('Le secret d'ennuyer est celui de tout dire.')

Our favourite Martians are here

Voltaire's aliens were friendly visitors and interested only in philosophical discourse with humans, but H.G. Wells' menacing Martians in *The War of the Worlds* invaded Earth to enslave humanity. At 8.50 pm on 30 October 1938, the day before Halloween, the marauding monsters (without their Halloween masks) arrived at Grovers Mill, a sleepy hamlet in New Jersey, at least on radio:

> The metal casing is definitely extraterrestrial – not found on this earth ... This thing is smooth and, as you can see, of cylindrical shape ...

Someone's crawling out of the hollow top … I can see … two luminous disks – are they eyes? It might be a face. It might be … [SHOUT OF AWE FROM THE CROWD]

Good heavens, something's wriggling out of the shadow like a gray snake. Now it's another one, another one, and another one! They look like tentacles to me. I can see the thing's body now. It's large, large as a bear and it glistens like wet leather. But that face, it – ladies and gentlemen, it's indescribable. I can hardly force myself to keep looking at it, so awful. The eyes are black and gleam like a serpent. The mouth is V-shaped with saliva dripping from its rimless lips that seem to quiver and pulsate. The monster or whatever it is can hardly move …

Ladies and gentlemen, I have a grave announcement to make. Incredible as it may seem … those strange beings who landed in the Jersey farmlands tonight are the vanguard of an invading army from the planet Mars.

Millions of listeners heard this 'live commentary' when they tuned in to a popular Sunday night programme on CBS radio, *The Mercury Theatre on the Air*, that featured plays directed by Orson Welles. The play that night was an adaptation of *The War of the Worlds*, presented as simulated news bulletins and scene broadcasts, presumably to heighten the dramatic effect. Welles had made another important change: he transplanted the action from H.G. Wells' Victorian England to contemporary America.

As the play unfolded, dance music from a hotel was interrupted several times by fake news flashes about a professor at an observatory noting a series of gas explosions on the planet Mars. Bulletins and scenes followed, reporting the landing of a 'meteor' in New Jersey, the discovery that

the 'meteor' was a 'metal cylinder' containing strange creatures from Mars armed with 'death rays', and that nearly
7,000 soldiers had been crushed and trampled to death.

The radio drama was introduced as a work of fiction,
but those who missed the announcement at the beginning
(the next one didn't arrive until 40 minutes into the play)
confused fiction for fact, and 'a wave of mass hysteria
seized thousands of radio listeners throughout the nation
between 8.15 and 9.30 o'clock,' the *New York Times* reported
next day in a lead story – 'RADIO LISTENERS IN PANIC, TAKING
WAR DRAMA AS FACT' – that filled nearly three-quarters of a
page of the broadsheet. 'The broadcast, which disrupted
households, interrupted religious services, created traffic
jams and clogged communications systems, was made by
Orson Welles, who as the radio character, "The Shadow,"
used to give "the creeps" to countless child listeners,' the
report said. 'This time at least a score of adults required
medical treatment for shock and hysteria.' Mass hysteria
mounted so high that in some cases people told the police
they saw the invasion.

The report recounted several examples of the panic,
including: 'In Indianapolis a woman ran into a church
screaming: "New York destroyed; it's the end of the world.
You might as well as go home to die. I just heard it on the
radio." Services were dismissed immediately.' And: 'The
bartender of a tavern … closed the place, sending away six
customers, in the middle of the broadcast to "rescue" his
wife and two children.' The broadcast even fooled two
Princeton University geologists: 'They armed themselves
with the necessary equipment and set out to find a speci-
men. All they found was a group of sightseers searching
like themselves for the meteor.'

The New York Times

Copyright, 1938, by The New York Times Company.

tered as Second-Class Matter,
Postoffice, New York, N. Y.

NEW YORK, MONDAY, OCTOBER 31, 1938.

Radio Listeners in Panic,
Taking War Drama as Fact

*Many Flee Homes to Escape 'Gas Raid From
Mars'—Phone Calls Swamp Police at
Broadcast of Wells Fantasy*

Figure 2. The front page of the New York Times, *31 October 1938, with
the lead story on a radio broadcast of a dramatisation of H.G. Wells'* The
War of the Worlds, *which sent thousands of Americans fleeing their homes
in panic.*

In 1877 Earth and Mars were in 'favourable opposition' –
that is, their orbits had brought them to their closest
together, something which takes place every 30 years or so.
During that year the American astronomer Asaph Hall discovered Phobos and Deimos, two small moons of Mars, but
the Italian astronomer Giovanni Schiaparelli did even better.
He noticed that the Martian surface was criss-crossed with
a network of about 100 lines. He saw them again in 1879
and again in 1881. He called these lines *canali* ('channels'),
which suggested that they were a natural feature. However,
the Italian word was mistranslated into English as 'canals'.
The idea of such large-scale artificial structures led to wild
speculations about life on Mars. The American astronomer
Percival Lowell became the champion of those who
believed that these canals were the work of intelligent
beings on the red planet. In 1905 he published a popular
book about the Martian civilisation, which fuelled the
public's fascination with Mars. (Carl Sagan once observed
that intelligence was certainly involved in the creation of

Martian canals. The only question was which end of the telescope the intelligence was on.)

When Wells wrote his novel in the late 1890s (it was published in 1898), he was very much inspired by the works of Lowell and other astronomers. He was also familiar with the historical debate about other worlds. The first edition of the novel has a prefatory quotation from Kepler: 'But who shall dwell in these worlds if they be inhabited? ... Are we or they Lord of the Worlds? ... And how are all things made for man?'

At least after *The War of the Worlds* we were the Lord of the Worlds: 'The Martians – *dead!* – slain by the putrefactive and disease bacteria ... after all man's devices had failed, by the humblest things that God, in his wisdom, has put on this earth.'

There are two film adaptations of the book: Byron Haskin directed the 1953 original; Steven Spielberg directed the 2005 remake.

Hoyle's frolic

In Fred Hoyle's science fiction novel *The Black Cloud*, the alien is one giant leap of imagination. It takes on the form of a cloud of interstellar gas – as large as the distance from the Sun to Earth and as massive as Jupiter – that can move as it wishes from star to star. This 500-million-year-old intelligent organism has a 'brain' that consists of neurological structures built from complicated molecules. The creature can increase the capacity of its brain by extending these structures and by learning to use its brain in the best way to solve problems as they arise.

Scientists notice the cloud when it approaches the solar system on a course that is predicted to bring it between the Sun and Earth, causing a global catastrophe. They discover that the cloud is alive and try to communicate with it. Once they establish contact, they warn the cloud that certain aggressive governments are trying to destroy it by hydrogen bomb rockets, an act 'sillier than trying to kill a rhino with a tooth-pick'. The cloud reverses the course of the rockets so that they 'hit the exact points they started from', destroying many cities. Eventually, the cloud decides to leave, but not before it has enjoyed Micromégas-like conversations with scientists and has expressed its views on matters ranging from the origin of headaches to the nature of intelligence. Some samples:

On why intelligent animals are unusual on planets: 'Living on the surface of a solid body, you are exposed to a strong gravitational force. This greatly limits the size to which your animals can grow and hence limits the scope of your neurological activity. It forces you to possess muscular structures to promote movement … Indeed your very largest animals have been mostly bone and muscles with very little brain … By and large, one only expects intelligent life to exist on a diffuse gaseous medium, not on planets at all.' On reproduction: 'I can live indefinitely, you see. Therefore I am not under the necessity, as you are, of generating some new individuals to take over after my death.'

In the book's preface, Hoyle says that he hopes that his scientific colleagues will enjoy this 'frolic'. He has also a disclaimer to the effect that none of the characters should be identified with the author. 'Be that as it may, the central character, Chris Kingsley, is a professor of astronomy at Cambridge, and Hoyle's handling is distinctly autobiographical,' says Simon Mitton in *Conflict in the Cosmos: Fred Hoyle's Life in Science* (2005). '*The Black Cloud* reads authentically because Hoyle sets the action in places he knew well: Caltech, Palomar Observatory, the RAS, and Cambridge.

His scientists are doing real science, not fantasy science: We see them using calculus, scientific diagrams, and the EDSAC-1 computer – all graphically described.'

The Black Cloud was Hoyle's first science fiction novel and was a sensational bestseller when published in 1957. It has now achieved classic status. Hoyle wrote and co-authored many other science fiction books before his death in 2001, but it is *The Black Cloud* for which he is chiefly remembered as a science fiction writer.

Hoyle, an astrophysicist, was a staunch supporter of the steady-state theory and never gave up his belief in it: the universe has no beginning and will have no end. This theory is now considered flawed and the Big Bang theory is widely accepted: the universe began when a single point of infinitely dense and infinitely hot matter exploded sponta-neously. The name 'Big Bang' was given by Hoyle. It was meant to be a put-down when he first used it scornfully in a radio talk in 1950.

A brooding ocean

The alien – an ocean that lives, thinks and acts – in Stanislaw Lem's science fiction novel *Solaris* is another giant leap of imagination. But unlike the black cloud, this ocean is not 'talkative'. Humans cannot communicate with it; they cannot even comprehend its intelligence.

The distant planet Solaris orbits two suns – a red sun and a blue sun. Its sole organic inhabitant is an ocean that covers the entire planet. The ocean expresses itself by elec-trical and magnetic impulses in a more or less mathematical language. These 'monologues' unfolding in the depths of the ocean's brain are incomprehensible to scientists who have been studying it from a space station orbiting the planet. They are convinced that they are confronted with 'a monstrous entity endowed with reason, a protoplasmic ocean-brain enveloping the entire planet and idling its time away in extravagant theoretical cogitation about the nature of the universe'.

The ocean regularly produces gigantic constructions in various shapes from a substance resembling a yeasty colloid. The scientists describe them in artificial, linguistically awkward terms such as 'tree-mountains', 'fungoids', 'symetriads', 'asymetriads' and 'vertebrids'. Were the constructions the 'taciturn' ocean's way of expressing itself?

'I only wanted to create a vision of a human encounter with something that certainly exists, in a mighty manner perhaps, but cannot be reduced to human concepts, ideas or images,' Lem explained on his website 41 years after the publication of his book. 'The ocean neither built nor created anything translatable into our language that could have been "explained in translation". Hence a description had to be replaced by analysis – obviously an impossible task – of the internal workings of the ocean's ego. This gave rise to *symetriads*, *asymetriads* and *mimoids* – strange semi-constructions scientists were unable to understand; they could only describe them in a mathematically meticulous manner.'

Kris Kelvin, the protagonist of *Solaris*, is a psychologist who arrives at the space station to determine whether research into Solaris should be terminated for want of progress. But he finds the space station all but deserted, its straggling crew seemingly haunted by hallucinations of 'phantoms' from their respective pasts. In the middle of the night, Kelvin himself is visited by an exact replica of his wife, Rheya, who had committed suicide after being estranged from him.

Was the ocean using these phantoms created from

neutrinos to pry into scientists' minds? Yes, says Lem, but it did it in a radical way. 'It penetrated the superficial established manners, conventions and methods of linguistic communication, and entered, in its own way, into the minds of the people of the Solaris station and revealed what was deeply hidden in each of them: a reprehensible guilt, a tragic event from the past suppressed by the memory, a secret and shameful desire.'

Beyond the issue of communicating with an alien life form, *Solaris* is also about whether contact should be made. 'We don't want to conquer the cosmos, we simply want to extend the boundaries of Earth to the frontiers of the cosmos,' sermonises one of the scientists based on the Solaris station. 'We think of ourselves as the Knights of the Holy Contact. This is another lie. We are only seeking Man. We have no need of other worlds. We need mirrors. We don't know what to do with other worlds.'

Lem, who died at 84 in 2006, is one of the greatest science fiction writers of all time. He wrote dozens of books, all in Polish, but *Solaris* is his masterpiece. Described as 'intellectual science fiction', it was first published in 1961 and has been translated into more than 40 languages. In 1972 the legendary Russian director Andrei Tarkovsky adapted *Solaris* into a film of the same title, which was described at the Cannes Film Festival as 'the most intelligent and insightful film in the history of science fiction movies'. Hollywood director Steven Soderbergh's *Solaris* (2002) is a well-crafted remake of the Russian classic.

An invitation to a Vegan banquet

In *Contact*, Carl Sagan creates a 'Knight of the Holy Contact'. Dr Ellie Arroway, Sagan's protagonist, receives a coded message in a radio signal: the Message is a plan for building a machine to transport five passengers to its benevolent sender on the star Vega, 26 light years away.

Arroway, 'a pretty woman with evident scientific competence who forcefully expresses her views', is not only faced with the scientific and technical problems of building the Machine; she has to deal with political ('who speaks for Earth?'), religious ('is the purpose of the Message divine or satanic?'), economic ('fears that building the Machine would ruin the world economy') and gender ('some would interrupt her or pretend not to hear her') issues.

There is 'narrow vocational disquiet' as well: 'There were other intelligent beings in the universe. We could communicate with them. They were probably older than we, possibly wiser ... So the specialists in every subject began to worry. Mathematicians worried about what elementary discoveries they might have missed. Religious leaders worried that Vegan values, however alien, would find ready adherents, especially among the uninstructed young ...'

And there are cultural subtleties to be taken care of. At the World Message Consortium, the Chinese delegate stresses: 'We have received an invitation. A very unusual invitation. Maybe it is to go to a banquet. The Earth has never been invited to a banquet before. It would be impolite to refuse.' Who would refuse an invitation to a sumptuous Vegan banquet?

Arroway accepts the invitation and the challenge of confronting religious opposition, political rivalry and scientific jealousy. The Machine is eventually built and makes a trip to Vega, but we never find out what the Vegans look like. But you hardly expect to see humanoid-like aliens in his only novel, because Sagan believed that our own forms were such an accident of numerous random evolutionary events that the odds of aliens looking humanoid were small.

Like Sagan's immensely successful *Cosmos* – the 1980 book and TV series – *Contact* makes science accessible. It's filled with scientific facts and fervour, which makes it utterly believable. *Contact* was published in 1985, and the film of the same name, directed by Robert Zemeckis, was released in 1997, a year after Sagan's death.

'We inhabit an insignificant planet of a hum-drum star lost in a galaxy tucked away in some forgotten corner of a universe in which there are far more galaxies than people,' Sagan said in *Cosmos*. 'We make our world significant by the courage of our questions, and by the depth of our answers.'

From glimpses of history and fiction of other worlds we now move to questions of life on this world.

CHAPTER 3

Life

The announcement at the Eagle pub

In 1944 Erwin Schrödinger, the celebrated physicist famous for his 'cat' (not a real moggy, but a thought-experiment), wrote a little book, *What Is Life?*. Stepping outside his field of expertise, he speculated that life's genetic information had to be compact enough to be stored in molecules in 'some kind of code-script'. These molecules, passed from parent to child, are 'the material carrier of life'.

Nobel Laureates Francis Crick and James Watson, crackers of life's code, have both acknowledged how this revolutionary idea inspired them. In 1953, in 'a few weeks of frenzied inspiration', as *Time* magazine put it, the two young and unknown scientists solved the secret of life at the Cavendish Laboratory in Cambridge. On 28 February, Crick walked into the Eagle pub in Cambridge and, as Watson later recalled, announced that 'we have found the secret of life'. That morning they had discovered the last piece of the puzzle that revealed the double helix structure of DNA. In short, DNA consists of a double helix of two strands coiled around each other. When the strands are uncoiled, they can produce two copies of the original. This unique structure explains how DNA stores genetic information and how it passes this information on to the next generation by making an identical copy of itself.

The DNA (deoxyribonucleic acid) molecule is like a twisted ladder. Each 'side of the ladder' is made up of chains of alternating sugar and phosphate units. 'Rungs' are made from pairs of four chemical compounds called bases: adenine (A), thymine (T), cytosine (C) and guanine (G). The bases always pair in a specific manner: A pairs with T, and C pairs with G. Thus there are only four combinations: A–T, C–G, T–A and G–C. The genetic code is the sequence of bases along the length of DNA. This code determines the order in which amino acids are linked together to form proteins.

The DNA resides in the nucleus of the cell. It instructs the cell to make proteins that control all the chemical processes in the cell. It does it by making RNA (ribonucleic acid), a close cousin of DNA. RNA is made up of the same bases as DNA except that the base U (uracil) replaces the base T. RNA serves as the blueprint for making proteins needed by the cell. The instructions on RNA are in the form of a code that consists of combinations of three bases available on DNA. Each combination represents an amino acid. Of the 64 combinations possible, 61 represent the twenty amino acids (as only twenty amino acids occur in the cells of all organisms alive today, in some cases several combinations refer to the same amino acid). The other three combinations act as 'full stops' in the coded information. The code can be written in either DNA triplets or the RNA copy of triplets. As an example, the triplet TTA in DNA (or CUU in RNA) instructs the cell to add the amino acid leucine.

The sequence of base pairs along the length of the strands is not the same in DNAs of different organisms. It is this difference in the sequence that makes one life form different from another. The human genome, the full DNA sequence of humans, has about 2.9 billion base pairs, which are wound into 24 distinct sausage-like bundles, or chromosomes. A gene is a segment of a chromosome. It is a length of DNA which has a complete code for one protein. All life forms have DNA.

Although it is difficult to define life, everyone agrees that all living things must have a system for storing and duplicating instructions about their structure. For life on Earth, DNA provides such a system. Its most remarkable feature is its genetic code, which is the same for all life forms that exist on this planet. And this feature makes the genetic code as old as life itself.

Indecent haste

The debris of the Big Bang that marked the beginning of the universe began to fly away from the explosion point, is still flying and will keep on flying indefinitely. Time begins at the Big Bang, which happened about 13 billion years ago.

Our planet was formed about 4.6 billion years ago from a ring of gas and dust around the young Sun. For nearly 700 million years the young Earth was subjected to intense bombardment by gigantic asteroids, the debris left over from the formation of the solar system. One such impact gouged a chunk out of Earth and formed the Moon. Life appeared about 3.8 billion years ago, as soon as the cosmic bombardment had ended. Some biologists say that life appeared 'fully formed, with almost indecent haste'.

The quick appearance of life on a young and inhospitable Earth suggests that life is easy to create and therefore it could have easily arisen on other planets in the galaxy.

Charles Lineweaver and Tamara Davis of the University of New South Wales have looked at the implications of the remarkably rapid beginning of life on Earth for the probability of its evolving elsewhere. To them it is like winning a lottery: 'The probability of winning a lottery can be inferred from how quickly a lottery winner has won.' If a gambler buys a ticket every day in a daily lottery and wins on the third day, it is likely, though not certain, that the chance of winning is 1 in 3. Similarly, 'we won soon after life became possible on Earth' points to there having been a good chance of life developing on other planets. The researchers say that they are 95 per cent certain that, given a billion years, the chance of life starting on an Earth-like planet is at least 1 in 3. On planets older than 1 billion years, the chance decreases to 13 per cent (about 1 in 8). They agree that it does not necessarily mean that life is common in the universe.

Fast or slow, how did life begin? Where did it begin? Was life's first incubator in the early oceans, in clay rocks, or deep-sea hydrothermal vents? First, an out-of-this-world scenario. Some scientists, small in number but highly vocal, believe that life did not begin on Earth at all, but arrived from outer space.

A 'wild and visionary' hypothesis

William Thomson (also known as Lord Kelvin) was the greatest physicist of his time. For 53 years he was professor at the University of Glasgow, but he was a failure as a lecturer and teacher. He was so preoccupied with his work, that if any new idea came to his mind while lecturing, he would digress and forget all about the topic of his lecture. Yet he liked to illustrate his lectures with demonstrations. Once, to explain a point, he brought an old muzzle-loader rifle and shot it at a pendulum. On another occasion he brought two eggs, apparently one boiled and one raw, to show the difference in how they behaved when spun. When he began the demonstration, he said smiling: 'Both boiled, gentlemen.' He had quickly discovered that some mischievous student had decided to play a practical joke on him by secretly boiling both eggs.

No, he was not the archetypal absent-minded professor; he had an extraordinarily clear mind and a powerful personality. 'He was certainly inspiring to students,' according to one of his contemporaries at the university. 'His enthusiasm was infectious.' He once said: 'Science is bound by the everlasting laws of honour to face fearlessly every problem that can be presented to it.' He himself fearlessly faced many problems of science.

In his presidential address at the 1871 annual meeting of the British Association for the Advancement of Science in Edinburgh, Thomson tackled the question of the origin of life on Earth.

All old men of science from Aristotle to Newton accepted without any serious question that life could arise instantaneously from living or non-living matter: maggots from decaying meat; caterpillars from leaves; frogs from slime. This view, called spontaneous generation, was first challenged in 1668 by Francesco Redi, an Italian physician and poet. Redi prepared eight flasks, each with a variety of meat in it: a dead snake, some fish and pieces of veal. He sealed four jars and left the other jars open to the air. After a few days he found that only the open jars bred maggots. The meat in the sealed jars was just as putrid, but without maggots. He now repeated the experiment, but this time covering four jars with gauze tops instead of sealing them. Air could enter the jar, but not flies. Again, maggots appeared in the open jars only. From these experiments Redi concluded that maggots were not formed by spontaneous generation but came from eggs laid by flies.

But even Redi's experiments could not shake people's belief in this age-old idea. It was the theories of Louis Pasteur and Charles Darwin that finally laid to rest the idea of spontaneous generation. In 1862 Pasteur performed a series of experiments to find the answer to 'the question of generation, so called spontaneous': 'Can matter organise itself? In other words, can beings come into the world without parents, without ancestors? Here is the question to resolve.' He resolved the question by declaring that all life comes from other life. In 1859 Darwin published his theory of evolution of all present-day species from simpler forms of life through a process of natural selection. Scientists did not readily accept Darwin's theory, but it renewed scientific debate on the subject of the origin of life.

In his address, Thomson said that science has brought a vast mass of inductive evidence against the hypothesis of spontaneous generation. 'Dead matter cannot become living without coming under the influence of matter previously alive. This seems to me as sure a teaching of science as the law of gravitation.' He then addressed the question of the origin of life on Earth: 'Hence and because we all confidently believe that there are at present, and have been from time immemorial, many worlds of life besides our own, we must regard it as probable in the highest degree that there are countless seed-bearing meteoritic stones moving about through space. If at the present instant no life existed upon earth, one such stone falling upon it might, by what we blindly call natural causes, lead to its becoming covered with vegetation ... The hypothesis that life originated on this earth through moss-grown fragments from the ruins of another world seems wild and visionary; all I maintain is that it is not unscientific.' His audience did not consider the idea scientific. They were sceptical of the notion of micro-organism-bearing meteorites flying in space and seeding different worlds.

However, at least one great man was impressed with Thomson's hypothesis. This admirer was Svante Arrhenius, the Swedish chemist who gave us the chemistry of ions, for which he won the 1903 Nobel Prize. Ironically, Thomson opposed the ionic theory when it was first proposed, saying that he could not understand anything which could not be translated into a mechanical model (for this reason he had also rejected Maxwell's theory of electro-magnetic waves – see page 144). Somehow, the idea of life from outer space did fit into his mechanical and mathematical universe.

Arrhenius was a versatile genius and investigated many scientific ideas beyond chemistry. In 1896 he was the first to recognise that carbon dioxide acts as a thermal blanket around the globe, thus creating the greenhouse effect. In 1906, in a book translated into English as *World in the Making* (1908), he suggested that bacterial spores and other dormant micro-organisms escaped from another planet where life already existed, travelled through space and finally landed on Earth and began to grow and develop. He called this process *panspermia* (Greek for 'all seeds').

Arrhenius did not explain how life originated on other planets. He just said that life is eternal. It has always been there, so the question of its origin does not arise. But he did support evolution: 'Life must always recommence from its very lowest type ... and must pass through all the stages of evolution from the single cell upward.'

The idea of panspermia fascinated scientists of the 19th century; however, it never became an accepted scientific idea. In recent times many reincarnations of panspermia have appeared.

Reincarnation 1: directed panspermia

In 1971 Francis Crick and fellow molecular biologist Leslie Orgel attended the first-ever international scientific meeting on 'Communication with Extraterrestrial Intelligence' in Soviet Armenia. At the conference banquet their host, the astronomer Viktor Ambartsumian, called on each guest to propose a toast, after which they were expected to down a glass of vodka. 'I vaguely remember ... my own toast to all extraterrestrial Armenians wherever they may be,' recalls Orgel. 'I also remember Francis Crick seeking relief from vodka, pouring a tumbler of what looked like water from a large jug, only to find it was more vodka. The young Russian student who came to help him told him not to worry, "the last Englishman to come to a party here had to be carried home".'

At that meeting, with or without the help of vodka, Crick and Orgel hit on the idea that perhaps life on Earth originated from micro-organisms sent here, on an unmanned spaceship, by an advanced civilisation elsewhere. 'We called our theory directed panspermia,' recalls Crick. 'Panspermia is the idea that micro-organisms drifted to the Earth through space and seeded all life on Earth. We used "directed" to imply that someone had deliberately sent the micro-organisms here in some way.'

Two years later, Crick and Orgel detailed their intriguing hypothesis in a paper in *Icarus*. They admitted that the chances of extraterrestrial micro-organisms reaching Earth either as spores driven by radiation pressure from another star or embedded in meteorites are extremely small. However, it is possible if someone decides to do it.

Are you and I the result of an experiment conducted aeons ago by drunken ET nerds?

Two biological anomalies support the hypothesis. The first concerns the universality of the genetic code. There is only one genetic code for all forms of life. Why is it so? No satisfactory answer from molecular biologists has come up yet. The universality is due to a 'seed' sent by an extra-terrestrial civilisation. As life originated from this 'seed', its genetic code replicated in all forms of life, resulting in a universal genetic code. That's the view of Crick and Orgel.

The second anomaly is the importance of molybdenum in biological systems. Many enzymes require the assistance of this trace element for their proper functioning. This would not have surprised molecular biologists if Earth were relatively rich in this metal, but it constitutes only 0.02 per cent of Earth's composition. Scientists expect the chemical make-up of living organisms to reflect to some extent the composition of the environment in which they evolved. 'If it could be shown that the elements represented in terrestrial living organisms correlate closely with those that are abundant in some class of – molybdenum stars, for example – we might look sympathetically at panspermia theories,' Crick and Orgel say.

They also discuss a number of questions likely to be raised against their hypothesis. Has there been enough time in the life of our galaxy for the sequential development of two advanced civilisations, one on Earth and the other on some planet beyond the solar system? Earth-like planets existed as much as 6.5 billion years before the formation of the solar system, and 3.8 billion years elapsed between the appearance of life on Earth (wherever it came from) and the development of our own technological society. Thus the time available makes it possible that technological societies existed elsewhere in the galaxy even before the formation of Earth.

The other important question is about the possibility of safe transport of life over interplanetary distances. Researches by other scientists have shown that life could probably be preserved for periods of more than a million years if suitably protected and maintained close to absolute zero. This time-frame supports Crick and Orgel's calculations, which suggest that at a speed of 96,000 kilometres per hour, a spaceship packed with micro-organisms – blue-green algae, for example, as it has simple nutritional requirements – could infect most planets in the galaxy. However, they warn strongly that under no circumstances should we risk infecting other planets.

Even if we heed this warning, we could accidentally 'infect' other planets in the way that our own planet might have been infected by micro-organisms in the garbage left over by extraterrestrial visitors. This preposterous idea was suggested in 1960 by maverick American astronomer Thomas Gold. He imagined that interstellar visitors forgot to clean up after having a picnic on Earth. Do piles of garbage left in picnic spots around the world hint that we have inherited this bad habit from our interstellar ancestors?

Crick revisited the notion of directed panspermia in his book *Life Itself: Its Origin and Nature* (1981). 'How *could* such stuff be considered seriously?' he asks. 'The whole idea stinks of UFOs or the Chariots of the Gods or other common forms of silliness.' He replies: 'The kindest thing to say about directed panspermia, then, is to concede it is indeed a valid scientific theory, but that as a theory it is premature.'

Reincarnation 2: lithopanspermia

Scientists have identified more than 130 molecules in interstellar space so far, including many organic molecules such as benzene, sugars and ethanol. However, amino acids have eluded them. Twice – in 1994 and 2002 – scientists announced that they had spotted glycine, the simplest of all amino acids, but both times the findings did not stand up to closer examination. Glycine is the holy grail of panspermia. 'Finding glycine is about 5 per cent of the way to proving the idea,' Fred Hoyle remarked in 1994 when he heard of the discovery. 'It's the right way, but it's only 5 per cent.'

Scientists ridiculed the idea that interstellar space contains organic molecules when Hoyle and Chandra Wickramasinghe advanced it in the 1970s. The common belief then was that there were no molecules in space except atomic hydrogen and ionic hydrogen. They also suggested that life with all its basic genetic information originated not on Earth but on a grand cosmic scale. Chemical building blocks of life are present in interstellar clouds. When these clouds collapse to form comets, they provide likely sites for the origin of life. Micro-organisms multiply inside a comet, which has a warm, liquid interior. An impact of a comet about 3.8 billion years ago could have led to the start of terrestrial life.

'We know as a matter of fact that comets do eject organic particles, typically at a rate of a million or more tonnes a day,' they say. 'This was what Comet Halley was observed to do in 1986. Comet Halley went on doing just that, expelling organic particles in great bursts, for almost as long as it remained within observational range. The infra-red spectrum of dust from Comet Halley matched precisely the laboratory spectrum of bacterial grains.' They believe that only the minutest fraction (less than one part in a trillion) of the interstellar bacteria needs to retain viability for panspermia to hold sway.

Like comets, meteorites can also distribute microorganisms (the distribution of microbes by comets or meteorites is called lithopanspermia). Even today, about 100 tonnes of debris from comets and meteorites arrives on Earth daily. This cosmic debris may bring in to the planet microbes responsible for diseases of plants and animals. 'The boldest answer must be yes; that is to say extraterrestrial biological invasions never stopped and continue today,' Hoyle and Wickramasinghe say. They hold cosmic debris responsible for many epidemics of a global nature, like the influenza of 1918 and the plague of Athens in 430 BC. They point out many anomalies in the distribution and spread of the 1918 and 1968 flu epidemics, and conclude that simple person-to-person spreading was not an adequate explanation, whereas the atmospheric dispersal of a space-borne agent was more convincing.

All this sounds plausible, but the question remains how bugs can survive the lethal radiation in space. The answer probably is that in interstellar clouds a thin layer of carbon material forms around micro-organisms, protecting them against damaging radiation.

The Murchison meteorite is the most famous meteorite to hit Australia. It fell near the town of Murchison, Victoria, in 1969, and was, in fact, a shower of hundreds of small pieces of carbon-bearing matter. In 1970 a team of American scientists analysed a few stones picked up soon after the fall and found several amino acids in the sample. The team also concluded that these amino acids seem to have been formed before the meteorite reached Earth. Ten years later, a piece of the meteorite came into the possession of German palaeontologist Hans Pflug. Upon close examination by an electron microscope, he noticed structures similar to fossilised remains of a terrestrial flower-like bacterium,

Pedomicrobium. Pflug refused to make any conclusion from his discovery: 'Either there is a fossil biomaterial in the meteorite or previous criteria used to identify microfossils in ancient terrestrial rocks are cast into doubt.' Some scientists have dismissed these similarities merely as 'mineralogical artefacts'.

Recent researches suggest that a galactic hitch-hike by micro-organisms is full of hazards but still feasible. German scientist Gerda Horneck has carried out a series of remote-controlled experiments aboard the Russian Foton satellite. She first released 50 million unprotected spores of a bacterium outside the satellite. Ultraviolet radiation from the Sun killed all the spores. She then released batches of 10,000 to 100,000 spores mixed with lumps of red sandstone one centimetre across. Nearly all the spores survived. The experiments suggest that meteorites as small as one centimetre across could carry life, if they complete their journey within a few years.

Calculations by H. Jay Melosh of the University of Arizona show that micro-organisms could even survive for millions of years in space if they are embedded in the interior of huge chunks of rock. This could happen when an asteroid impact ejects rocks from a planet. It seems that the impacts that produced craters on Earth greater than 100 kilometres across would each have ejected millions of tonnes of rocks carrying micro-organisms into interplanetary space, much of it in the form of boulders large enough to shield those micro-organisms from radiation. 'Although terrestrial organisms in these rocks would have the opportunity to colonise the planet,' Melosh says, 'it seems unlikely they would find the conditions suitable for propagation.' It also seems unlikely that scientists will soon accept Hoyle and Wickramasinghe's hypothesis.

Reincarnation 3: radiopanspermia

When a star's radiation – not comets or meteorites – propels micro-organisms into interstellar space, it is called radiopanspermia. Radiation from stars can exert enormous pressure on nearby matter. In 1928 British astrophysicist Arthur Eddington showed that, in order to remain stable, the inward gravitational force of a star must balance the outward radiation pressure. Stars of more than 60 solar masses cannot exist, because they are torn apart by radiation pressure.

Solar radiation regularly ejects dust grains from the solar system and they can travel through space to other star systems. Any bacteria and viruses embedded in these dust grains usually do not survive because of ultraviolet rays. However, if a radiation-absorbing material like carbon shields the micro-organisms, there are chances that they could reach other star systems. Therefore, it is also possible that micro-organisms could have arrived from another solar system in our galaxy to Earth by radiation pressure.

Jeff Seeker of Washington State University and his colleagues have studied the chances of survival against harmful ultraviolet radiation of bacteria and viruses ejected from a solar system like ours. They conclude that 'the effect of vacuum and low temperatures are not as bad as might be expected, and it is the ultraviolet radiation of the host stars that presents the greatest biological hazard'. If bacteria and viruses are encapsulated in small protective mantles of material, roughly half a micrometre thick, it is possible to screen them from lethal radiation. There is always some damage to bacterial and viral DNA and RNA, but this damage is reparable if the bacteria and viruses find a hospitable environment at the end of their journey. Viruses also need a specific living host for repair. (Viruses consist of a small number of genes encased in a protective coat of protein. They are not organisms; they can grow and multiply only when inside living cells.)

The protective layer increases the mass of micro-organisms and therefore considerable radiation is required to eject them into space. Stars known as red giants can provide such intense radiation. In about 4.6 billion years, when the Sun exhausts its supply of hydrogen, it will also turn into a red giant, 250 times its present diameter and thousands of times more luminous. Red giants are old stars, and their planets are most likely to harbour life. These stars emit very little ultraviolet radiation, but they are bright enough to carry micro-organisms into space. Calculations by Seeker's team show that particles with the size of about one micrometre would take about 35 years to leave a one-solar-mass red-giant star system and 1 million years to reach a neighbouring star system. Beyond that time, interstellar radiation becomes a serious threat. To travel further, micro-organisms would need to hop on from planet to planet: establish life on a new planet, which in time would cast out new micro-organisms, and so on.

This red-hot idea cannot be proved until live micro-organisms are found in interstellar space. Says Paul Parsons, a UK science writer: 'If the theory is correct, the notion of life existing in isolated pockets in the galaxy is a misleading one. It becomes more appropriate to think of a ubiquitous "galactic life", of which life on Earth is just a small part.'

Reincarnation 4: Mars–Earth panspermia

Ever since Galileo caught sight of Mars through his newly-built telescope in 1609 (he described it as a 'spherical body illuminated by the Sun'), the red planet has been a source of intrigue and imagination.

It has inspired scores of (almost all B) movies: *Invaders from Mars*, 1953, 1986 (Martians brainwash residents of a small town); *Mars Needs Women*, 1968 (Martians invade Earth to capture women; almost proves the theory of panspermia, remarks a wit); *The Alpha Incident*, 1984 (deadly virus from Mars reaches Earth in a space probe); *Total Recall*, 1990 (Arnold Schwarzenegger travels to Mars); *My Favourite Martian*, 1999, adapted from the 1960s TV show (a friendly Martian visits Earth).

And its orange-red face has launched more than 30 space missions (at least eighteen of which have experienced some kind of failure): *Mariner 4*, the first spacecraft to fly past Mars on 14 July 1965; *Mariner 9*, the first spacecraft to orbit Mars on 13 November 1971; *Viking 1*, the first spacecraft to land on Mars, 20 July 1996 (it scooped some topsoil but failed to show any convincing sign of organic molecules); *Mars Exploration Rover Spirit* completed one Martian year, or almost two Earth years, on Mars on 21 November 2005 (no news yet of the presence of life); *Astrobiology Field Laboratory*, planned for launch during the next decade, will conduct a robotic search for life.

Mars movies take it for granted that the planet is teeming with life, but Mars missions have a defining question: Is there any evidence of life in the planet's past? If so, could any of these microscopic creatures still exist today?

The answer is not simple, as Mars has had a chequered past which is laid bare in the barren, bone-dry, freezing desert we see today: sinuous river valleys, mountainside gullies, shorelines and flood plains all sculpted long ago by flowing water, and perhaps by icy glaciers formed by polar snow. The climate was warmer, the atmosphere thicker to trap heat from the Sun. Why has Mars changed so dramatically?

The three main theories of what has transformed Mars 'read like the plot of a detective novel', according to Ralph Lorenz of the University of Arizona. Perhaps Mars was 'murdered', suffering a slow, drawn-out death by suffocation as impacts from asteroids and comets eroded the atmosphere; or perhaps it was death by 'suicide', a case where Martian silicate rocks reacted with atmospheric carbon dioxide to form carbonate minerals: 'If there was no other source of carbon dioxide, the atmosphere would gradually have been sucked into the surface of the planet.' Or maybe the planet died of 'natural causes' that resulted when, gripped in a severe ice age, its carbon dioxide atmosphere condensed to form permafrost at the poles or beneath the entire surface. The permafrost would be hidden by the dust that covers Mars. The American space agency NASA's Mars Polar Lander, launched on 3 January 1999, was expected to provide clues to Lorenz's Mars mystery by digging water ice with a robotic arm. It was lost on arrival exactly eleven months later. The mystery remains unsolved.

The mystery of life on Mars was almost unravelled on 7 August 1996 when NASA announced in a historic press conference that a primitive form of microscopic life might have existed there about 3.6 billion years ago, when Mars was about 1 billion years old. NASA scientists had found evidence of fossilised, microscopic life in an ancient Martian meteorite, known as ALH84001. The meteorite was catapulted away from Mars 15 million years ago when a huge asteroid hit the surface. After drifting through interplanetary space for millions of years, it landed on Earth about 13,000 years ago. The potato-sized meteorite was found in 1984 in Allan Hills, a jagged region of ice in Antarctica.

Four features of the meteorite which, when taken together as a whole, suggest ancient life on Mars: 1) the carbonate patterns form a unique signature of life, and the density and composition of the patterns is consistent with how terrestrial bacteria operate; 2) polycyclic aromatic hydrocarbons, organic compounds usually created by bacteria, are present in the meteorite; 3) tear-shaped magnetite globules present in the meteorite are created by bacteria on Earth; and 4) the presence of microscopic fossil-like structures, the most dramatic evidence of all.

The NASA announcement filled metres of newspaper columns and hours of radio and TV time around the world. The London *Times* (8 August 1996) screamed on its front page, CLINTON HAILS DISCOVERY OF LIFE ON MARS, and quoted the President as saying: 'It speaks to us across billions of years and millions of miles.' The editorial in the same edition was more poetic. It quoted the 17th-century French mathematician and philosopher Blaise Pascal – 'the eternal silence of these infinite spaces terrifies me' – and hoped that 'Pascal's infinite spaces may contain an infinite number of other civilisations, sufficient to satisfy every taste'. The *New York Times* (8 August 1996), however, was circumspect in its editorial: 'For the moment, confirmation is surely needed before we let this inconclusive finding propel us too far toward an intensified hunt for life on other worlds … it would be prudent to hold the jokes.'

As it turned out, the joke was on NASA. Many scientists were then sceptical of NASA's claim ('at best premature and more probably wrong'), and most scientists now agree that Martian 'microfossils' in ALH84001 are not indicative of life. ALH84001 (and 30 or so other known meteorites of

Martian origin) may or may not be messengers of past Martian life, but no one challenges the idea that microscopic life could travel from Mars to Earth, or the reverse, on a meteorite.

Recent researches by American scientists David Warmflash and Benjamin Weiss suggest that microorganisms could have survived a journey from Mars to Earth. Meteorite evidence shows that material has been transferred between planets throughout the history of the solar system and that this process still occurs regularly. Could biological entities (such as RNAs, DNAs, living cells or bacteria) survive the journey? The journey has three stages:

Ejection from the parent planet: Laboratory experiments show that certain bacteria can survive the accelerations and jerks that they would experience during a typical high-pressure ejection from Mars. Calculations show that a small percentage of ejected rocks could indeed leave Mars without any heating.

Travel through interplanetary space: Biological materials have better chances of surviving the perils of interplanetary space – exposure to ultraviolet rays, X-rays and gamma rays – if carried inside meteorites.

Entry into the Earth's atmosphere: During entry through the Earth's atmosphere, heat rays reach only a few millimetres at most into the meteorite's interior, so organisms buried deep in the rock would certainly survive. Organisms embedded in interplanetary dust would avoid heating, as dust particles decelerate gently in the upper atmosphere.

In short, science does not reject the idea that life on Earth could have arisen from an extraterrestrial seed, probably from Mars.

In the (primordial) soup

The eminent Scottish biochemist and geneticist, J.B.S. Haldane, would have dismissed the idea of an extraterrestrial seed as 'worthless nonsense'. For two reasons:

First, he once said (jokingly) that the normal process of acceptance of a scientific idea has four stages: i) this is worthless nonsense; ii) this is an interesting, but perverse, point of view; iii) this is true, but quite unimportant; iv) I always said so.

Second, he was a firm believer in the idea that life originated on Earth. His ideas have helped shape the current picture of the origin of life.

Darwin avoided the question in his classic *The Origin of Species* (1859), except in the final paragraph when he said that 'the Creator' originally breathed life 'into a few forms or into one', and then it evolved 'from so simple a beginning' into 'endless forms most beautiful and most wonderful'. However, in a letter written in 1871 to his botanist friend Joseph Hooker, he suggested that life could have arisen, without supernatural intervention, 'in some warm little pond, with all sorts of ammonia and phosphoric salts, light, heat, electricity, etc. present', where it was possible 'that a proteine compound was chemically formed'. This compound was 'ready to undergo still more complex changes; at the present day such matter would be instantly devoured or absorbed, which would not have been the case before living creatures were formed'. For the next century, Darwin's 'private hypothesis' dominated thinking on the subject.

In 1924 Aleksander Oparin, a 30-year-old Russian biochemist, presented the first viable theory of the origin of life by natural physical and chemical means here on Earth: in the early atmosphere, simple inorganic compounds combined to form complex organic compounds, which formed the first living cell.

During Stalin's reign, Oparin held a powerful position in the Soviet Academy of Sciences. Rumour has it that he used to sit down for dinner with a bottle of cognac on one side and a bottle of vodka on the other, both of which would be empty by the end of the meal. Oparin, who died in 1980, is admired not for his drinking prowess but for the 'primordial soup' he concocted.

He suggested that early in the Earth's history the atmosphere was rich in hydrogen. Simple inorganic hydrogen compounds such as water, methane and ammonia could form organic compounds. Gradually these organic compounds fell from the atmosphere to the ground, where the rain – which occurred when Earth cooled and water vapour condensed – washed them into pools and ultimately into oceans. Over millions of years the organic molecules in this 'primordial soup' joined together into long chains of proteins and DNA molecules until a cell appeared which possessed the right kind of reactions and right kind of compounds to be considered an organism. This first cell could replicate itself and therefore it filled the bill for the first living organism.

In 1929 Haldane independently proposed the same general theory: 'The first living and half-living things were probably large molecules synthesised under the influence of the Sun's radiation, and only capable of reproduction in the particularly favourable medium in which they originated. Each presumably required a variety of highly specialised molecules before it could reproduce itself, and it depended on chance for a supply of them. This is the case today with most viruses ... which can grow only in presence of the complicated assortment of molecules found in a living cell.'

In 1953 Stanley Miller, a young graduate student working in the laboratory of Nobel Prize-winning chemist Harold Urey at the University of Chicago, provided the first experimental support to the 'primordial soup' theory. He subjected a mixture of methane, ammonia, water vapour and hydrogen to a series of electrical charges. He imagined this to be a rough duplication of conditions on the primitive Earth when the primordial soup was subjected to bolts of lightning. After a week, the inorganic molecules had joined to form several amino acids, building blocks for proteins which make up cells. Urey was exultant: 'If God didn't do it this way, He missed a good bet.' (The Murchison meteorite described earlier was also shown to contain a number of the same amino acids that Miller identified.)

Miller's 'primordial soup' has been the staple diet of biology textbooks for decades, but is now under serious threat. 'Although the taste of the soup seems quite agreeable now – at least with its broad experimental base – there are critics who go for other courses as far as the origin of life is concerned,' says Günter von Kiedrowski of the University of Ruhr. A taster of these new courses:

Crêpes on rocks: It has been easy to create amino acids and other pre-biotic molecules in the laboratory, but the next step – linking these molecules to form longer, self-replicating molecules – has proved challenging. These pre-biotic molecules rarely link into chains of more than twenty, but in order to replicate, chains of about 60 molecules are necessary. That's the minimum number required to write a primitive genetic code. Recent experiments have now shown that long, chain-like molecules up to 55 units long can assemble themselves on clay surfaces instead of in solution. The researchers used a clay called montmorillonite, which is formed from weathered volcanic ash and is familiar in many households as cat litter. These experiments, says Kiedrowski, suggest that 'the polymers of life were more likely to have been baked like pre-biotic crêpes than cooked in a pre-biotic soup'. 'The common message,' he adds, 'is that the earliest forms of life may have proliferated by spreading on surfaces.'

Scientists have found another property of montmorillonite clay of possible relevance to the origin of life: it makes droplets of fat molecules rearrange themselves into small bubbles, similar to the membranes that make up the walls of living cells.

Crystals in clay: The idea that clay particles may have greatly facilitated the emergence of the first cells is not new. It was first popularised by Graham Cairns-Smith, a Scottish chemist, in the 1960s. He thinks that life may have originated in small, flat clay crystals. These crystals exhibit basic qualifications for life: they can assemble themselves, carry information, reproduce by duplicating their own structures, and thus pass along information from generation to generation. Clay crystals repeat the same molecular pattern over and over again. When a new crystal is formed, it retains the original pattern, including the defects in the pattern of the old crystal, rather like a mutation in a strand of DNA. The defects in patterns can encode information and therefore serve as simplistic memories. Cairns-Smith suggests that clay crystals could have served as molecular moulds that incorporated life's building-blocks and organised them in precise arrays.

Fool's gold: Günter Wächtershäuser is a German patent attorney who also holds a doctorate in chemistry. His work has appeared in prestigious journals. He suggests that when we think of the origin of life we must not think of the DNA, the RNA or the cell. These things must have come later. 'Life didn't begin with a soup of chemicals, which cannot do anything, which are inactive,' he says. 'It's change that leads to life.' Natural chemical reactions provide that change. Life must have started in the simplest possible way, as a cycle of natural chemical reaction that repeated itself. This cycle most likely started on the surface of pyrite, the shiny crystal known as fool's gold. It catalyses energy-production reactions known as the citric acid cycle that occurs in all organisms, including us. Wächtershäuser describes these reactions as a 'primitive metabolism': 'It's my theory that life began with a metabolism, and this metabolism invented everything else – genetic machinery and cells that enclose it.'

All this did not necessarily happen in Oparin's primitive oceans. Wächtershäuser claims that this process can happen anywhere: 'In vents under the ocean, in volcanoes on land, even deeper down, anywhere where gases come out from Earth. In my theory, the origin of life goes on today.'

The origin of life, according to Thomas Gold

Thomas Gold, the brilliant scientific gadfly who died in 2004, is remembered for his unconventional theories. 'Gold's theories are always original, always brilliant, usually controversial – and usually right,' writes the renowned physicist Freeman Dyson in the preface of Gold's last book,

The Deep Hot Biosphere (1999). In this book, Gold states the case for his most controversial theory – he calls it his 'heretical views' – that contradicts the conventional wisdom that petroleum and natural gas are fossil fuels, the fossilised remains of plants and animals that died millions of years of ago. On the contrary, says Gold, these resources are constantly being manufactured deep in the Earth by natural processes from the initial materials that formed Earth. He calls his theory, which ensures a virtually inexhaustible supply of petroleum and natural gas, 'deep-earth gas' theory. He supplements this theory with the concept that a 'deep hot biosphere' has been flourishing on these deep resources for billions of years.

Extremophiles are microbes that thrive under conditions that would kill other creatures – in deep-sea hydrothermal vents, rock chimneys that grow above volcanic vents in the sea floor, through which erupt hot, mineral-rich fluids; inside rocks buried kilometres below the Earth's surface where there is no oxygen, no organic food; in frozen Antarctic sea water; or in acidic, alkaline or saline environments. Extremophiles offer the most probable model for testing the hypothesis that life exists on other planets. The reasoning is simple: if life can exist in extreme environments on Earth, it can also exist in the extreme environments of other worlds; for example, beneath the icy surface of Jupiter's moon Europa.

Recent discoveries of extremophiles far deeper in the Earth's crust than previously believed support Gold's idea of a deep hot biosphere which covers the entire crust down to a depth of several kilometres. Unlike the surface life, which is fed by photosynthesis (the process by which plants, algae and certain bacteria convert sunlight to chemical energy), the deep life is fed directly by chemical energy.

Both life forms have the same genetic system. This means that they have a common origin. Where did life begin – on the surface or at depth?

For Gold there is no question that life began inside the Earth and then migrated to the surface. He supports his theory with considerable evidence, which includes:

Radiation: Rocks can protect life at depth from harsh radiation. Surface life forms have evolved protective coatings and pigments to ward off harmful radiation, but this protection was not available to the very first life forms. Subsurface conditions offer a more agreeable situation for the beginning of life.

Energy: The interior of the Earth can provide abundant chemical energy, which is in short supply on the surface. The Sun is the main source of energy on the surface. The first living cell could not have performed the feat of photosynthesis. A large chemical supply on the surface will begin only when organisms have developed complex photosynthesis reactions. Photosynthesis probably followed when some microbes found it advantageous to live at or near the surface to enrich their energy sources by the use of sunlight.

Temperature suitability: With the exception of active volcanoes, the temperature gradient (the rate at which temperature changes) below the Earth's surface holds steady at any given depth; there is no 'weather' and no glacial episodes or boiling water. This is in sharp contrast to conditions at the surface – particularly on land – where enormous temperature shifts may occur seasonally and daily.

Carbon copies of LUCA

Everything alive is made up of cells (living cells were first described in 1665 by the English scientist Robert Hooke, who named them so because they reminded him of tiny monks' rooms in monasteries). Some bacteria have only one cell; some animals millions.

The last universal common ancestor (LUCA) of modern cells, from which all organisms descended, lived about 3.8 billion years ago. Although LUCA was not necessarily the first living organism, hyperthermophiles, extremophiles that live at temperatures over 80 degrees Celsius in deep-sea hydrothermal vents, are probably its closest living relatives. When LUCA was born, the planet's atmosphere lacked oxygen, and abundant minerals in hydrothermal vents provided the energy source needed for sustaining life. Hyperthermophiles are the most ancient living forms known. Even if they are LUCA's closest relatives, it does not necessarily follow, as claimed by Gold, that life itself originated in deep-sea hydrothermal vents.

As there are no fossil records of LUCA, no other known 'footprints', LUCA's identity has proved elusive. Molecular biologists are trying to create a portrait of the mother of life by comparing the genes of all life forms. Even this method has flaws. Biologists estimate that a simple LUCA might have had as few as 450 genes, and its diameter was about 300 nanometres. A recent study of genomes of 100 species has found only 60 genes in common. Apparently, genes have been lost from species' genomes as organisms adapted to new conditions.

While scientists are still searching for LUCA's identity, they are sure that it had three things that are present in the cells of all living organisms today: DNA for genetic code, RNA for retrieving the genetic code for making amino acids, and ribosomes, molecular factories for assembling proteins from amino acids.

Here is a breakdown of the components of cells and the chemical elements present in them: *water*, universal solvent (hydrogen, oxygen); *protein*, involved in all functions (carbon, hydrogen, oxygen, nitrogen, sulphur); *fat*, energy storage (carbon, hydrogen); *carbohydrate*, cell walls (carbon, hydrogen, oxygen); *DNA*, *RNA* and a tiny molecule called *ATP*, genes and energy (carbon, hydrogen, oxygen, nitrogen, phosphorus).

As far as elements are concerned, cells consist mostly of carbon, hydrogen, oxygen and nitrogen, together with small quantities of sulphur and phosphorous. Plants and animals also require numerous other trace elements (small concentrations of elements present in a sample) such as sodium, potassium, magnesium, iron, zinc, calcium, chlorine, fluorine and copper. Of the 92 elements present in nature, only 21 play a role in life.

Carbon, hydrogen, oxygen and nitrogen are the prime elements of life. Why is life made of these elements and not others? The answer is simple: because they are four of the five most abundant elements on Earth. The fifth is helium, but it's an inert gas and takes absolutely no part in any known chemical reactions.

The story of life as we know it is the story of water and carbon. Without them there would be no life on our planet, but could there be life without water or carbon (or both) on other planets? That we do not know – yet.

Water is a miracle that makes our world possible. It has many amazing characteristics that help to support and sustain life:

Supersolvent: The ability of water to dissolve almost everything enables it to carry nutrients through the bodies of plants and animals.

Ability to climb against gravity: Water can move easily through extremely narrow tubes and ooze through invisibly tiny holes. This 'capillary action' lifts water up from under the ground, through the soil to the roots of plants. It then ascends through stems and leaves.

A hearty appetite for heat: Water can absorb more heat than any other substance without considerable rise in temperature. The slow cooling and warming of water prevents extreme climatic changes and protects living things from the shock of abrupt temperature changes.

Ice is lighter than water: Practically every substance contracts as it becomes colder, but water – when cooled below 4 degrees Celsius – expands to have about 10 per cent more volume as a solid than as a liquid. If ice did not float, oceans and bodies of water would be frozen from the bottom up and there would be no living things in them.

The peculiar behaviour of water is the result of the so-called hydrogen bond between molecules of water. The hydrogen bond links the positively-charged nucleus of a hydrogen atom in one water molecule to the negatively-charged electron cloud of a nearby oxygen atom in another water molecule. There is another property of water that makes it unique. It is a polar molecule, which simply means that one part of the molecule has a positive charge and the other negative charge. Other polar molecules can dissolve in water but non-polar molecules cannot. The walls of cells are made of non-polar carbohydrates, ensuring that they will not dissolve in water.

Some scientists have suggested that liquids such as ammonia or methane – Saturn's moon Titan has oceans of methane – could replace water in life forms evolved on other planets. Two- to four-centimetre-long centipede-like worms have been found living in mushroom-shaped mounds of frozen methane seeping up from the floor of the Gulf of Mexico.

Life began when organic molecules, molecules containing carbon, slowly began to assemble in liquid water. Carbon has an extraordinary ability to form compounds with other elements. A carbon atom can form bonds with four other atoms. These atoms may be other carbon atoms or non-metal atoms, especially hydrogen, oxygen, nitrogen, sulphur and phosphorus. Most organic compounds are combinations of carbon with one or more of these five atoms. Carbon atoms link together to form straight chains or rings of usually five to eight atoms. Some organic molecules contain as many as 100,000 atoms.

Carbon-based molecules in terrestrial life have two limitations: a) they cannot obtain liquid water essential for their well-being below freezing point; and b) they start breaking down above a few hundred degrees Celsius. This narrow range of temperatures makes them suitable for life on Earth only.

Besides carbon and water, life also favours left-handedness. Each amino acid, except glycine, exists in mirror-image left- and right-handed forms. But all the naturally-occurring proteins in all organisms are made up of left-handed amino acids. Why this bias towards left-handedness? There are two explanations: a) amino acids that fall to Earth from space are more left-handed than right-handed because space radiation destroys more right-handed amino acids (mostly left-handed amino acids have been found in meteorites); and b) left-handed amino acids are more stable in water and that's why proteins are made up of them (life, after all, started in water). The first explanation boosts the case of panspermia supporters; the second one deflates it.

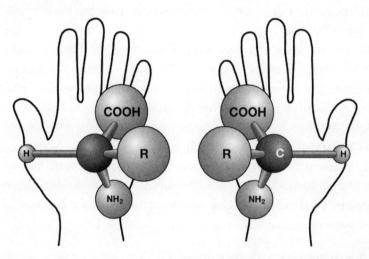

Figure 3. Amino acids exist in two mirror-image forms. (Illustration: NASA)

Silicon also has some life-giving properties of carbon: it can form long chains to which other elements bind, and it is abundant in the universe (sand is silicon dioxide), though not as abundant as carbon. These peculiarities of silicon have led some scientists to suggest that there might be silicon-based life on other planets. However, Arrhenius was totally against this speculation. He said in 1908: 'All organic beings in the whole universe should be related to one another, and should consist of cells which are built up of carbon, hydrogen, oxygen and nitrogen. The imagined evidence of living beings in other worlds in whose constitution carbon is replaced by silicon or titanium must be relegated to the realm of improbability.' Carl Sagan, 'a carbon chauvinist' in his own words, said in 1976: 'Carbon compounds are not just more abundant but more stable. So while life on other planets would probably not look like life on Earth because its internal biochemistry would be astoundingly different, I think it would be based on carbon.'

While scientists root for carbon, science fiction writers relish the idea of silicon-based life. Even the gifted H.G. Wells was 'startled by the visions of silicon-aluminium organisms … wandering through an atmosphere of gaseous sulphur'. More recently, in 2267 in fact, *Star Trek*'s Commander Spock has encountered silicon-based life, the Horta, on the planet Janus VI.

Some carbon-based descendants of LUCA may find it difficult to understand the babblings of the sand-blobs of Janus VI. They may also question the idea of Earth as a super-organism.

A super-organism called Gaia

If there were no life on Earth today, its atmosphere would be like that of Mars, with a lot of carbon dioxide and no oxygen. Life on Earth did not adapt to the conditions it found, but has helped to change the environment to make it hospitable; for example, nearly all the carbon dioxide, nitrogen and oxygen in the air today come from biological sources.

The Gaia hypothesis seeks to explain the interdependence of life and the physical world: the Earth's temperature and atmospheric composition are regulated by feedback between living things and the physical world, and these feedback mechanisms have evolved with the biosphere over the past 3.8 billion years.

James Lovelock, a maverick British scientist, first proposed the Gaia hypothesis in 1972. He said that life does not exist on Earth only because material conditions happen to be just right. Life on Earth defines the material conditions needed for its survival and makes sure that they stay there. The Earth's living matter, air, oceans and land surface form part of a giant system that is able to control temperature, the composition of the air and sea, the pH of the soil and other parameters so as to be optimum for survival of the biosphere. If the system is damaged dangerously, it can repair itself.

The system seems to exhibit the behaviour of a single organism, even a living creature. Lovelock's neighbour, novelist William Golding (*Lord of the Flies*, 1954), suggested the name Gaia for the hypothesis. Gaia was the name given by the ancient Greeks to their Earth goddess.

Lovelock sees Earth as a living organism of which we are a part; not the owner, nor the tenant, nor even a passenger on that obsolete metaphor, 'Spaceship Earth'. Any species that adversely affects the environment is doomed, but life goes on. Years ago, an article on Gaia in *New Scientist* magazine led a reader to wonder: 'Could AIDS be the response of the ecosystem to a rogue species, *Homo sapiens*, perceived as a threat to the stability – or indeed the survival – of the system?' This Gaian thought generated a controversy in the letters pages of the magazine. However, the last word came from a cartoon with the caption: 'If Gaia meant us to die of AIDS she would not have made rubber trees.'

When first proposed, the Gaia hypothesis was not well received: most scientists either ignored it or criticised it as untestable. Since then the hypothesis has gained much scientific respectability, and some even accord it the status of a theory. A theory is tested by experiments or by making some predictions from it.

One of the predictions of the Gaia hypothesis is that there was never any life on Mars – or, if there was life in its early history, it should now exist in some form. The hypothesis maintains that life quickly gains control over the environment and regulates it. Mars was much warmer once than it is now, and if life existed on Mars in its early history, the slow cooling should not have wiped it all out. If the Gaia hypothesis is correct, then life should exist on Mars now.

The original Gaia hypothesis – organisms regulate the planet for their own advantage – has gone through a series of adjustments and now includes a new scientific approach, geophysiology, that views life and the physical world as part of a global system capable of environmental regulation. Geophysiology describes organisms interacting among themselves and with their environment at a global scale, and recognising that these interactions lead to self-regulation, one can consider the system a super-organism. You may call this super-organism Gaia.

An international team of geologists led by Minik Rosing of the University of Copenhagen has recently claimed that photosynthesis might have arisen 3.8 billion years ago, about 1 billion years earlier than has previously been shown. The appearance of photosynthesis coincides with the emergence of continents. Earth's three main layers (the core, mantle and crust), the oceans and atmosphere took shape in the first 200 million years of its 4.6-billion-year history, but the stable continents did not form until 3.8 billion years ago.

The oldest areas of continental crust can still be seen in Acasta, Canada. The Acasta crust was formed from granite, a rock not found elsewhere in the solar system. Granite is created when volcanic rocks melt and react with minerals in earth and water. Granite rocks are lighter than volcanic rocks, so they rise to the surface, forming a stable crust – the continental shelf. The puzzle is not how the granite formed, but why it took so long, as the ingredients were there well before the Acasta rocks.

Rosing's team believes that photosynthetic life is the missing piece in the puzzle. With the evolution of photo-synthetic life, solar energy was channelled into the geochemical cycle. This additional energy kick-started chemical changes in Earth itself. Life, it seems, created continents. The researchers' claim, if generally accepted, would support the Gaia hypothesis that life helps to create conditions it needs to survive.

Beyond Gaia – the selfish biocosm

Life not only plays a role in shaping our planet, it also plays a role in shaping the universe. The cosmos is remarkably hospitable to life. It's a living creature, a biocosm. It's so finely tuned to life that if any of its fundamental physical laws and constants – from the 'just right' strength of the Big Bang that produced it to the relative strengths of gravity, electromagnetism and the so-called strong and weak nuclear forces – were slightly different, there wouldn't be any life. The laws and constants are intelligently designed to yield life and even more competent intelligence. Why is the universe so life-friendly?

Life and intelligence have not emerged in a series of random events, but essentially are hard-wired into the cycle of cosmic creation, evolution, death and rebirth. This hard-wiring was done by super-intelligent extraterrestrials to give birth to life and intelligence. Life and intelligence, in turn, are the means used by the universe to reproduce itself by creating 'baby universes' which themselves, in turn, are able to create more 'baby universes'.

No, this is not from the blurb of a new science fiction book, but a summary of a new hypothesis, called Selfish Biocosm, which has won praise, if not approval, from many scientists. James Gardner, an American complexity theorist, science essayist and lawyer, presents it in his book *Biocosm* (2003). The essence of the hypothesis, according to Gardner, which he himself labels 'speculative', is that 'the universe we are privileged to inhabit is literally in the process of transforming itself from inanimate to animate matter'.

He agrees that, to be acceptable as a genuine scientific hypothesis, it should be testable. He presents four possible tests: a) the discovery of extraterrestrial intelligence would prove the hypothesis; b) the proof of evolution of sentience in non-primate species such as dolphins would support the hypothesis by indicating that the emergence of intelligence is a relatively robust, species-neutral phenomenon; c) the creation of artificial life and its ability to evolve would support the hypothesis; and d) the emergence of trans-human machine intelligence, the combined human and machine intelligence, is necessary for the feats of cosmological replication.

It will all happen when life on Earth becomes super-intelligent like the extraterrestrial super-intelligent beings that created it. Gardner believes that humans and their probable progeny are the evolutionary predecessors of the supremely evolved intelligence that will emerge in the distant future. At the end of the cosmic evolutionary cycle, a super-intelligent entity is sufficiently advanced to be capable of engineering, or re-engineering, the basic laws and constants of physics. Let's find out whether the present most intelligent life form on Earth will live long enough to become one day the super-intelligent 'architects of the universe'.

ETs will always have someone to talk to when they visit Earth

American astrophysicist J. Richard Gott is an expert on black holes, regions of space where the gravity is so strong that nothing, not even light, can escape. He also loves using the 'Copernicus principle' to forecast the future duration of events as mundane as a couple's current relationship or as momentous as the longevity of the human race.

Copernicus rejected the prevalent belief that Earth stood still at the centre of the universe. The Copernicus principle says that one's location is unlikely to be special, or to put it bluntly, wherever or whenever we are, it's nothing special.

The maths of the Copernicus principle, according to Gott, works as follows: 'If there's nothing special about your observation of something, then there's a 95 per cent chance that you are seeing it during the middle 95 per cent of its observable lifetime, rather than during the first or last 2.5 per cent. At one extreme, the future is only $\frac{1}{39}$ as long as the past. At the other, it is 39 times as long. With 95 per cent certainty, this fixes the future longevity of whatever you observe as being between $\frac{1}{39}$ and 39 times as long as its past.'

In 1969 Gott visited the Berlin Wall, which was then eight years old. He applied the above formula (with 50 per cent likelihood; in a later version he changed this value to a greater certainty of 95 per cent), and predicted that it would last more than $2\frac{2}{3}$ years but less than 24. The Wall came down twenty years later in 1989.

Before you say 'wow', let's apply Gott's maths to *Homo sapiens*, who appeared about 200,000 years ago. It means that we'll last at least another 5,100 years but less than 7.8 million years. Turn the music on – there's no need for gloom and doom, yet. And life on Earth, which appeared about 3.8 billion years ago, will survive between 97 million and 148 billion years. We may have gone, but ETs will always have someone to talk to when they visit Earth.

Right now we have only one example of life. One example could be a fluke, says John A. Ball – 'with two examples we can do statistics'. Even if there were one other world with life, the right statistics would be: one plus one does not mean two, but many worlds with life.

The statistics of Gott's theory is stark when we apply his theory to his own theory. It was first published in 1993; therefore, it should have a future life of between two months and 156 years before it's sucked into a black hole.

CHAPTER 4

Worlds

6, 7, 8, Pluto ...

Five wandering points of light in the night sky are easily visible with the naked eye, and were known to the ancients – the Babylonians, the Egyptians, the Chinese, the Hindus and others. The Greeks called them *planētēs* ('wanderers'). Today we know the five as Mercury, Venus, Mars, Jupiter and Saturn – all named after Roman gods.

Even the inquiring Aristotle did not know that he was standing on the sixth planet – Earth. This ignorance continued until 1543 when Copernicus declared that 'Earth was simply one wanderer of many'. In 1609, after learning of its invention in Holland, Galileo designed and built his own telescope. With his twenty-power telescope, he was the first to see the wandering points of light in the night sky as spheres – as other worlds like Earth. In 1633 Galileo removed any remaining doubts about the motion of Earth from the minds of adherents of Aristotle when, after recanting his belief in the Copernican system at his trial by the Inquisition, he stamped his foot and muttered inaudibly: 'Eppur si muove.' ('And it does move.') Many doubt that such a protestation ever happened, but that kick, real or imagined, has kept Earth in perfect orbit ever since.

The Sun and its six planets were now a complete heavenly family, the solar system. No one, not even William Herschel who discovered the seventh planet in 1781, imagined that there could be another planet beyond Saturn.

Herschel was born in Germany, but moved to England when he was nineteen and lived there for the rest of his life. He was a talented musician and for many years worked as an organist at the Octagon Chapel in Bath. He became interested in astronomy when he was 35. A year later, in 1774, he saw the stars for the first time through a telescope, a telescope that he had built himself. He gradually gave up his musical career and became obsessed with constructing better telescopes and watching the sky through them. 'He was incessantly making fresh mirrors, or trying new lenses, or combinations of lenses to act as eye-pieces, or projecting alterations in the mounting by which the telescope was supported,' remarks Robert Ball in his rich biography of Herschel in *The Great Astronomers* (1895). His younger sister Caroline, who at this time had come to live with him and look after the housekeeping, complained that in his 'astronomical ardour he sometimes omitted to take off, before going into his workshop, the beautiful lace ruffles which he wore while conducting a concert, and consequently they became soiled with the pitch employed in the polish of his mirrors'.

In the history of science, Herschel holds the position of the greatest astronomer of his time, Caroline a grander position: the first important woman astronomer. Caroline not only managed household duties, she also mastered the arduous art of polishing mirrors and the complex mathematics of astronomical calculations. When Herschel was at the telescope, from dusk till dawn, 'Caroline sat by him at her desk, pen in hand, ready to write down the notes of the observations as they fell from her brother's lips'. As the telescope was in the open air, during some winter nights the ink would actually freeze in her pen. Next day she would carefully transcribe the observations during the night, and make necessary calculations so that her brother could observe the next swathe of the sky.

On 13 March 1781, while systematically observing stars in the constellation of Gemini, William Herschel came across an object that appeared more than a little point of light, as stars did in his 7-inch-aperture reflecting telescope. When he applied a higher magnifying power, the object appeared like a tiny yellow-green disc. He thought he had discovered a comet and reported it as such. He continued his observations and calculations for months, which showed that the object had a circular orbit, like a planet's, rather than the elliptical orbit of a comet. 'The organist at the Octagon Chapel at Bath had, therefore, discovered a new planet with his home-made telescope,' as Ball recounts the momentous, first recorded discovery of a planet in human history.

Herschel, an unknown astronomer until then, was now a celebrity. He decided to name the planet 'Georgium Sidus' ('George's star') in honour of George III, King of England. The name was used in Britain for many years, but the planet was eventually named Uranus, after the father of Saturn in Roman mythology. King George gave Herschel and Caroline the titles of 'The King's Astronomer' and 'Assistant to the King's Astronomer' and life-long pensions. Herschel – and Caroline – continued to scan the skies until his death in 1822. Their discoveries include two moons of Uranus, two new moons of Saturn and numerous double stars, nebulae and clusters.

The discovery of Uranus posed a problem for astronomers. They kept on finding it in the wrong parts of the sky; the planet was apparently drifting away from its predicted orbit. Was there an unknown planet beyond Uranus pulling it out of its orbit?

Urbain Le Verrier, a French mathematician, assuming that an unknown planet was the culprit, decided to calculate its size and position. He published his first results in June 1846. After more work, on 18 September he wrote to Johann Galle at the Berlin Observatory to ask him to look for it. Galle received the letter on 23 September, and on the same night he spotted a small blue disc very close to the position where Le Verrier had predicted it. He wrote to Le Verrier on 25 September: 'Monsieur, the planet of which you indicated the position really exists ... My God in heaven it is a big fellow.'

On the other side of the Channel, *The Times* announced the discovery on 1 October with the headline: DISCOVERY OF LE VERRIER'S PLANET. It noted that some astronomers in London had also observed it in the evening of 30 September. The newspaper, however, did not know that the British also had a stake in the discovery. A young mathematician, John Couch Adams, also suspected that an undiscovered planet was affecting Uranus' orbit. In October 1845 he sent his predictions to the Astronomer Royal, George Airy, who wrote back in November asking a minor technical question. Airy did not show any enthusiasm for Adams' calculations until July 1846, when he heard of Le Verrier's predictions. He met with James Challis, director of the Cambridge Observatory, and John Herschel, William's son and a prominent astronomer, and asked Challis to begin a search at the Observatory. However, he was doubtful of success, as he wrote later: 'It was a novel thing to undertake observations in reliance upon merely theoretical deductions; and while much labour was certain, success appeared very doubtful.'

On 10 September Herschel addressed a meeting of the British Association for the Advancement of Science in Southampton, in which he said that he saw the new planet as 'Columbus saw America from the shores of Spain. Its movements have been felt, trembling along the far-reaching line of our analysis.'

After reading the *Times* report, Airy prepared a set of documents in which he outlined Britain's case. Before these documents were published at the 13 November meeting of the Royal Astronomical Society, on 3 October Herschel wrote to London's *Athenaeum* magazine making public the role of Adams. Support for the British claim also came from Wilhelm Struve, director of the observatory at Polkowa, near St Petersburg. He wrote in the *Athenaeum* (20 February 1847): 'But impartial history will, in future, make honourable mention also of the name of Mr Adams, and recognise two individuals as having, independently of one another, discovered the planet beyond Uranus ... But Mr Adams's labours were unsuccessful because the two astronomers (Mr Challis of Cambridge and Mr Airy of Greenwich) to whom they were known hesitated to admit them without further examination.'

French astronomers were sceptical of the British claim; nevertheless, they agreed to name Le Verrier and Adams the co-discoverers. British astronomers accepted the name Neptune, after the Greek god of the sea, suggested by Le Verrier.

Airy's account had always rankled some science historians and writers. In 1964, prolific science writer Isaac Asimov described him as 'a conceited, envious, small-

Figure 4: A cartoon by Amédée de Noé, a French caricaturist and lithographer who became famous as Cham, published in the French news weekly L'Illustration (7 November 1846) at the height of the controversy over the discovery of Neptune. The cartoon shows a caricature of Adams looking through a telescope across the Channel at Le Verrier's notes.

minded man [who] ran the observatory like a petty tyrant'. In the mid-1960s, when historians looked for Airy's file at the Greenwich Observatory, they found it missing. It turned up in 1998 at the Chilean Institute of Astronomy. More documents have appeared since.

After analysing all the documents, a trio of international historians – William Sheehan, Nicholas Kollerstrom and Craig Waff – concluded that though Adams did some interesting work, he never calculated the exact position of the new planet. In *Scientific American* (December 2004) they accuse Britain of stealing Neptune: 'Now the passions stirred by the international rivalries of the 1840s have died down … we can affirm that Adams does not deserve equal credit with Le Verrier for the discovery of Neptune … The achievement was Le Verrier's alone.'

After a century and a half, Neptune has been restored to its rightful discoverer, but it has yet to complete one orbit since its discovery: it takes Neptune about 165 years to orbit the Sun.

Even after Neptune's discovery, Uranus kept on misbehaving. The existence of Neptune had explained irregularities in its orbit – almost, but not quite. Percival Lowell, the wealthy American astronomer who had earlier championed the idea that intelligent Martians had created canals all over the surface of Mars, suggested that an undiscovered planet was affecting Uranus and Neptune. He called it Planet X (X as in unknown). In 1895 he started searching for it from the Lowell Observatory near Flagstaff in Arizona, which he had built a year earlier, partly for the very purpose of finding the new planet. The technique for the search for new planets had changed since the discovery of Neptune. Astronomers now made two photographs of the same part of the sky, several weeks apart. Using an optical device called a blink comparator, they then manually checked both photographs, made up of millions of dots, hoping to find a dot that had moved from its place from the earlier photograph (computers now do this painstaking task). Lowell spent ten years searching through countless photographs, but he found nothing. A disappointed man, he died in 1916. However, his observatory continues to operate.

In 1929 Clyde Tombaugh, an earnest high school graduate who was too poor to afford college, wrote to the Lowell Observatory, enclosing drawings of his observations of Mars and Jupiter through his home-made 9-inch-aperture telescope. Impressed with the quality of his work, the Observatory offered him a job working with their new 13-inch photographic telescope. After less than a year's searching and photographing the sky, on 18 February 1930 he saw a faint point of light moving a bit from night to night. The 24-year-old self-taught astronomer had found the ninth planet. It was named Pluto for the Greek god of the underworld (Mickey Mouse's dog and the new artificial element plutonium were subsequently named after the new planet).

But was it Planet X, the planet that was pulling Uranus and Neptune off course? No. Pluto is so tiny that it has no appreciable effect on the motion of either Uranus or Neptune. In the 1990s astronomers recalculated the masses of Jupiter, Saturn, Uranus and Neptune, using data from Voyager fly-bys, and concluded that Uranus and Neptune were right on course. Le Verrier, Lowell and other astronomers were wrong. There is no Planet X affecting Uranus and Neptune.

In 2003, American astronomers Michael Brown, Chadwick Trujillo and David Rabinowitz discovered an object which they thought was the tenth planet. Like Pluto, it's a rocky ball covered with frozen methane. Its 2,400-kilometre diameter makes it 110 kilometres larger than Pluto, and 10,300 kilometres smaller than Earth.

The new object, Eris, lies about three times as far from the Sun as Pluto, which makes it the furthest object ever seen in the solar system. Pluto and Eris both reside in the Kuiper belt of primordial icy bodies that surrounds the eight planets closest to the Sun. The belt is named after Dutch astronomer Gerard Kuiper who, in 1951, suggested that a belt of icy bodies might lie beyond Neptune. The Kuiper belt is more massive than the asteroid belt between Mars and Jupiter, and contains about 100,000 bodies larger than 100 kilometres across.

Except for Pluto, no one could find any objects in the Kuiper belt until 1992, when astronomers discovered the first one, an object smaller and more distant than Pluto. The discoverers wanted to name it Smiley, but the International Astronomical Union (IAU) gave it a boring catalogue number, 1992 QB1. Since this discovery, hundreds of objects like QB1 have been found. And that's just the tip of the iceberg, astronomers say. They also believe that many objects larger than Eris – possibly as big as Mars – are hidden in the Kuiper belt. More new worlds are waiting to be discovered at the edge of the solar system. It is highly unlikely that any of these frozen balls could harbour any form of life.

Pluto's problem

Within days of the discovery of Pluto, some astronomers expressed doubts about its identity. They said it was not a planet but a large, icy asteroid or a comet. The publicity put forward by Lowell Observatory highlighted that it was indeed Percival Lowell's Planet X. As no American had discovered a planet before, the news captured the imagination of the public. Intense media coverage and widespread public interest in suggesting a name – Pluto was suggested by eleven-year-old Venetia Burney from Oxford, England – drowned out the voices of dissent. The IAU accepted Pluto as the ninth planet.

Estimates of Pluto's size have been shrinking since its discovery. At the time, the Lowell Observatory announced that the new planet was at least as large as Earth. Later estimates suggested that it was half that size. In 1976 it shrank to a diameter of 3,500 kilometres. In 1978 astronomers discovered that Pluto has a relatively large moon, Charon, with a diameter of 1,180 kilometres. New measurements showed that Pluto was only 2,290 kilometres across, smaller than Earth's moon.

Even the discovery of Charon did not help Pluto's case. The existence of a moon is not a winning argument for calling a body a planet. Mercury and Venus do not have moons, but many other solar system bodies do have them. We now know that many asteroids have moons around them; for example, Ida, a 56-kilometre-long potato-shaped asteroid, has a 1.6-kilometre-long tiny moon called Dactyl. About 10 per cent of the Kuiper belt objects have moons. Eris also has a moon; it has been nicknamed Gabrielle, after Eris's travelling companion, until the IAU decides on an official name.

Besides its puny size, Pluto has many other peculiarities that make it an oddball among planets. It isn't a gas giant like its closest neighbours – Jupiter, Saturn, Uranus and Neptune. It isn't rocky like the so-called terrestrial planets – Mercury, Venus, Earth and Mars. Its composition, a mixture of ice and rock, resembles that of a comet. Its highly elliptical and oddly angled orbit – tilted 17 degrees relative to the plane in which other planets move – is also very unusual. No other planet has such a comet-like orbit. For about twenty years, Pluto's 248-year orbit brings it closer to the Sun than Neptune. It then becomes the eighth planet. The last such mix-up happened from 1979 to 1999.

So what is Pluto? A planet, a dormant comet or a Kuiper belt object? The bottom line is that if Pluto were discovered today, it would not be called a planet. In 1999 the IAU considered changing Pluto's planetary status, but decided not to go ahead after a public outcry led to hundreds of e-mail protests.

What's a planet anyway?

After the discovery of Eris in 2003, the *Wattle Review* website reported that an American man has added fuel to a furious astronomical debate by officially applying to the IAU for status as a planet. 'There isn't a good definition of what constitutes a planet,' says Gerald Finkelschmidt. 'I have put on few pounds these past couple of years, but I'm clearly below the limiting mass for thermonuclear fusion ... And I orbit the Sun, obviously. So, technically I qualify.' At that time the IAU's 'working' definition of a planet stated in part that planets were objects with masses below the limiting mass for thermonuclear fusion of deuterium, and that orbit stars. Whatever it meant, it obviously worked in favour of Finkelschmidt, who had also noted in his application that the atmosphere around him was high in methane, 'especially after I make a Taco Bell run'. The *Wattle Review* is a satirical website, so fictional Finkelschmidt's application is still going round and round in circles.

On a serious note, what is really a planet? Dictionaries tell us that a planet is a large, round heavenly body that orbits a star and shines by the star's reflected light. This simple definition is good enough for you and me, but not for astronomers. A brief look at some of their ideas:

Planets are round: Alan Stern, an American planetary scientist, suggests that the first criterion for planethood should be roundness. If a body is big enough, its gravity will conquer geological and mechanical forces and will pull it into a spherical shape. The lower limit for diameter is about 800 kilometres. Smaller objects – both asteroids and comets – are not round. For lawyers out there who want to quibble on the definition of roundness, Gibor Basri of the University of California at Berkeley warns that rotating planets are actually oblate; that is, they are flattened at the poles.

Planets orbit around stars in their own zones: The star, according to Basri, is an object so massive that its interior burns with thermonuclear fusion. It may be a single star or a multiple-star system. A planet orbits around a star, but it orbits alone in its zone. This excludes asteroids, which swarm in an orbit between Mars and Jupiter, and Kuiper belt objects, which orbit in a much bigger swarm. The criteria of roundness and orbit will add to the present roll-call of nine planets at least one asteroid, Ceres (diameter 930 kilometres), and three known Kuiper belt objects: Eris, Quaoar (diameter 1,300 kilometres) and Sedna (diameter 1,800 kilometres). Many other Kuiper belt objects yet to be discovered may also qualify.

Planets are less massive than thirteen times the mass of Jupiter: At about thirteen times the mass of Jupiter, the intense gravity of a body makes it hot and dense enough to start thermonuclear fusion. True stars are usually 70 or more times the mass of Jupiter. Any object more massive than thirteen Jupiters, Basri says, would be a 'brown dwarf' star, and certainly not a planet.

Farewell, Pluto

In 2006 the IAU ended the controversy about the definition of a planet once and for all by declaring that 'a "planet" is a celestial body that a) is in orbit around the sun, b) has sufficient mass for its self-gravity to overcome rigid force so that it assumes a hydrostatic equilibrium (nearly round) shape, and c) has cleared the neighbourhood around its orbit'. This means that the solar system consists of eight planets: Mercury, Venus, Earth, Mars, Jupiter, Saturn, Uranus and Neptune.

Pluto and Eris fail to qualify as planets as they meet criteria a) and b) but not c). The IAU has created a new category of objects, called 'dwarf planets', for Pluto and Eris-like objects. This new category also includes Ceres. The IAU expects to classify more Kuiper belt objects as 'dwarf planets' in the coming years.

'It is scientifically the right thing to do, and is a great step forward in astronomy', says Michael Brown, co-discoverer of Eris, who was disappointed to learn that Eris would not be the tenth planet. As far as Pluto is concerned, it will be a long time before the ninth planet disappears from people's consciousness.

Pluto has been a long-standing myth that is difficult to kill. This myth has been part of our culture and consciousness in countless different ways, from textbooks (almost all), to websites (including NASA's), to poems ('Here are nine planets we know/Round and round the Sun they go'), to mnemonics for remembering the name and order of the nine planets ('My Very Energetic Mother Just Served Us Nine Pizzas'), to posters and museum exhibits (with one notable exception – for many years, Pluto has been on ice at the solar system exhibits at the American Museum of Natural History in New York).

Planets around faraway stars

Ever since the 18th century, when German philosopher Immanuel Kant and French mathematician Pierre Simon de Laplace put forward their nebular hypothesis for the formation of the solar system – a spinning cloud of gas and dust broke into rings that condensed to form the Sun and the planets – the possibility of planets orbiting other stars has fascinated scientists and science-fiction writers alike. The writers simply let their spaceship travel faster than light to exotic planets in the Milky Way, but scientists had to find a planet before they could even dream of travelling to it.

The search for extra-solar planets started in earnest in 1981 when a team of Canadian astronomers led by Gordon Walker observed dozens of Sun-like stars. Their decade-long search yielded nothing. It appeared that, after all, we are alone. At least it did until October 1995, when the news of the confirmation of a planet orbiting another star hit the headlines around the world (A PLANET ORBITING A SUNLIKE STAR CHALLENGES NOTIONS THAT EARTH IS UNIQUE – *New York Times*).

The planet, discovered by Michel Mayor and Didier Queloz of Geneva University, orbits 51 Pegasi, a Sun-like star only 44 light years from the Sun. It's a Jupiter-like gas giant 140 times the mass of Earth, and so close to its star that it whips around it in a mere four days.

It seems that Mayor and Queloz opened the floodgates. Astronomers have since discovered more than 200 extra-solar planets, many of them in multi-planet systems. All of these planets are at least fifteen times the mass of Earth, except the one that orbits a red dwarf 22,000 light years away. Discovered in 2005, it is the smallest planet detected outside the solar system. Its mass is around five times that of Earth, and it takes about ten years to orbit its star at about 2.6 times the distance of Earth to the Sun. It's as frigid as Pluto, with a surface temperature of about –222 degrees Celsius, probably too cold to support life.

The importance of this discovery, by Jean-Philippe Beaulieu and his team of the Institut d'Astrophysique de Paris, lies in the fact that this rocky, Earth-like planet orbits a red dwarf. Red dwarfs are up to 50 times fainter than the Sun, and make up 70 per cent of the stars in the Milky Way. Does it mean that Earth-like planets are abundant?

Discovering extra-solar planets, especially smaller planets, is extremely difficult. We cannot see an extra-solar planet directly, even with the most powerful telescopes. Planets reflect light from the stars they orbit. The reflected light is millions of times dimmer than the light from the star. As planets are close to their star, it's difficult to distinguish this dim light from the star's bright light. Planet-hunters use various indirect techniques.

The most common is the 'wobble technique'. Here's how it works. A planet orbiting around a star is like two whirling dancers pulling each other. As a planet circles its star, its gravitational pull on the star causes the star to wobble back and forth. The amount of this tiny wobble depends on the mass of the planet; the heavier the planet, the more the star wobbles. This wobble creates a 'Doppler shift' in the spectrum of the star as seen from Earth.

If you have noticed the changing pitch of the whistle of a train as it runs past you standing on a platform, you know what the Doppler shift is. The pitch is higher when the train is approaching you, and lower when it's moving away from you. The effect was described in 1842 by Austrian physicist Christian Doppler, and is caused by the change in frequency of a sound or light wave with reference to the observer as the source moves away from or towards the observer. Three years later, Dutch meteorologist Christoph Buy Ballot tested the effect for sound waves in an endearing experiment – as a moving source of sound, he used an orchestra of trumpeters standing in the open car of a railroad train, whizzing through the Dutch countryside. In 1848, French physicist Armand Fizeau showed that the Doppler shift applies to light coming from distant stars. No, he didn't use a spaceship full of trumpeters to test his idea. If a star is moving away from us, the light it emits shifts slightly towards the red end of the spectrum, which makes it appear slightly redder. If a star is moving closer to us, it appears slightly bluer. This phenomenon is called the red shift.

The spectrum of the wobbling star can reveal details such as the mass and the orbit of the planet, but not its nature. The wobble technique works well for massive Jupiter-like planets, but it fails to detect lighter Earth-like planets.

Another planet-hunting technique is gravitational microlensing, which can pick up smaller planets. Microlensing is based on Einstein's theory of general relativity, which says that the mass of an object causes space to curve. As a result, light can be bent by gravity. When an object passes between a star and Earth, its gravity causes the light to bend towards Earth. The light is magnified or 'lensed' and the star seems brighter, for a few hours, days or weeks. This spark of brightness makes planet-hunters beam from ear to ear; it is the tell-tale sign of the presence of a planet.

Brave new worlds

What makes a planet fit for life? It must not be too hot, nor too cold. A planet's distance from its star determines whether it will be hot or cold. It must be at the right distance from its star for the existence of the elixir of carbon-based life – liquid water. It must not be too big, nor too small. If it is too big, its gravity will attract gases from space. It will then be like Jupiter, which has an outer shell of hydrogen and helium. If it is too small, like the Moon, its pull of gravity will be so weak that it will not be able to hold on to oceans and an atmosphere, as water or gases will be lost in space. It must also be relatively safe from space hazards: bombardment by asteroids and comets, and blasts of dangerous radiation from space.

For an intergalactic estate agent our planet is prime real estate. It has the three 'must have' things in real estate: location, location and location. 'Any extraterrestrial civilization seeking a new world would place our solar system on their home-shopping list,' informs *Scientific American* (October 2001). Pray that old copies of *Scientific American* are not found in bug-eyed alien doctors' consulting rooms.

Earth not only has the right location, it has a solid surface that can support water. This surface has a unique feature – the movement of the planetary crust across the surface of the planet. Called plate tectonics, it provides many other life-supporting features. The Earth's crust is like a cracked egg shell. It is broken up into six large pieces and several smaller ones. The plates – as the pieces are called – are 100 to 150 kilometres thick and carry the continents and oceans on their backs like giant rafts. The plates are drifting about 2 centimetres per year (the rate at which our fingernails grow) over a layer of red hot melted rocks that underlie the rigid crust. If these drifting plates collide, they cause mountains to rise. If they move away from each other, new oceans are formed. These changes take place over millions of years. Most volcanoes and mountain ranges are concentrated along the plate boundaries (or where the boundaries used to be). No other 'terrestrial' planet has a linear mountain chain or shows evidence of plate tectonics. However, magnetic maps of Mars suggest that plate tectonics may have occurred on the red planet in its early history.

Atmospheric gases, mainly carbon dioxide and water vapour, form a blanket around Earth and keep it at a comfortable temperature for life. Without this blanket, Earth's temperature would be below freezing point. Volcanoes are critical in releasing carbon dioxide into the atmosphere. Plate tectonics recycles chemicals crucial to keeping the level of carbon dioxide in the atmosphere uniform, and it also makes possible the Earth's magnetic field, which shields us from the solar wind (a spray of charged particles from the Sun) and other lethal cosmic radiation. If plate tectonics stopped today, in a few million years the mountains would erode and our world would return to its state of 4 billion years ago: the entire planet covered by a global ocean, except for the odd volcano poking out.

The simplest forms of life take in water, carbon dioxide and nitrogen as nutrients and then release oxygen in the atmosphere. If we can find a planet with water, carbon dioxide and oxygen (or ozone, a form of oxygen usually found in the upper atmosphere) we will be one step closer to finding life – not necessarily complex life – on other worlds.

Where can we find a planet fit for life? In a habitable zone, of course. A habitable zone, also known as a Goldilocks zone ('Ahhh, this porridge is just right'), is a region around a star in which a planet's surface temperature is 'just right' for liquid water to exist. The Sun's habitable zone extends from after Venus to just before Mars, so only our planet is within this zone. These are the present limits of the zone.

But during the early history of the Sun, when it emitted less energy, Venus and Mars were probably also within the larger habitable zone. Perhaps water existed on these two planets in their early history. As the Sun heated up, Venus – with its thick carbon dioxide and sulphur dioxide atmosphere, and lying closer to the Sun – also heated up and lost its ability to hold liquid water. However, the geological history of Mars, which has very little atmospheric carbon dioxide, shows that it actually cooled off and also lost its ability to hold liquid water. Only the planet in the middle remained 'just right' like Goldilocks' porridge.

Europa, Jupiter's second moon, is an enigmatic world that defies the concept of a habitable zone. NASA's *Galileo* spacecraft, which orbited Jupiter from 1995 to 2003, found evidence of a deep ocean of liquid water under Europa's icy crust, which is a chilly –145 degrees Celsius. For us earthlings, water equals life. Is this ocean teeming with aliens? Stay tuned. NASA has been planning a spacecraft to orbit Europa in 2015.

Our solar system is, of course, within the galactic habitable zone, but how far does the zone extend? A team of astronomers led by Charles Lineweaver of the University of New South Wales has identified a ring that measures 21,000 to 27,000 light years from the centre of the galaxy as the galactic habitable zone. This region contains about 10 per cent of the galaxy's 250 billion stars, which were formed between 4 and 6 billion years ago. Because a parent star and its planets were formed from the same cloud of gas and dust, the star must have the right 'metallicity'; that is, enough heavy elements (elements heavier than hydrogen and helium) to sustain life.

The right metallicity is not the only prerequisite for life. The star must have lived through relative cosmic calm – not too many supernova explosions – for at least 4 billion years, the time it took for life to evolve on Earth. This prerequisite rules out dangerous inner zones of the galaxy from the habitable zone. 'If there are aliens, 75 per cent of them will have had longer time to evolve than we have,' Lineweaver says. 'This may be the most fundamental take-home message of this study.'

American astrophysicist William Danchi and French colleagues Bruno Lopez and Jean Schneider suggest that planet-hunters should look for planets around dying red giant stars. In a few billion years, when our Sun will become a red giant star, its habitable zone will lie beyond Uranus. Planets that are at present very cold and icy would warm up and could kick-start life. The researchers' analysis shows that the period over which these conditions change is very long – long enough for life to form. It took about 700 million years for life to emerge on Earth. They suggest that meteorites from a planet could transport micro-organisms from where life is ending to a planet where favourable conditions for its rebirth are emerging.

There are nearly 150 sub-giant and giant red stars within 100 light years of Earth (compared to about 1,000 Sun-like stars). It narrows down the search area for planet-hunters. As the search for other planets is also, in part, a search for life beyond Earth, we may be closer to the holy grail of astrobiology, the study of the likelihood of extraterrestrial life.

This story of brave new worlds has a punch-line, and it comes from Freeman Dyson, who believes in the possibility that life elsewhere is already adapted to living in a vacuum. Asteroids and comets, where gravity is weaker and moving from one world to another is easier, are the most likely habitats for life adapted to a vacuum. A planet would be a death-trap for such forms of life, he says. 'A planet for them would be like a deep well full of water for a human child.' He suggests that the swarm of smaller objects in the Kuiper belt in our solar system – and Kuiper belts in other star systems – have much more real estate, which provides more habitable surface area than all planets put together. 'I am willing to bet even money that when the first alien life is found it will not be on a planet,' he says. 'This is a bet which I will be happy either to win or lose.'

Before you dig into your pocket to place your bet, let's look at what other scientists have to say about the possibility of intelligent life on other worlds.

CHAPTER 5

Hypotheses

Making the bishop's wife happy

After life appeared on Earth, evolution took over. The main tenet of evolution, as proposed by Darwin in his monumental book, *On the Origin of Species by Means of Natural Selection* (1859), is that all present-day species have evolved slowly, over millions of years, from simpler life forms through natural selection, which favours individuals that are better adapted for survival and reproduction.

Did you know about the reaction of the bishop's wife to the suggestion that man is derived from the apes? 'Let's hope that it is not true – or if it is, that it won't become generally known.' This is an example of the furore caused by Darwin's theory, which was in complete contrast with the religious notion of the creation of Earth, its life forms and the rest of the universe by a divine being. To the annoyance of the bishop's wife, the idea of evolution has become generally known. But it still continues to generate enormous social and scientific debate.

After about 3.8 billion years of evolution, humans are at its pinnacle. Intelligence has helped them to get there. Intelligence is difficult to define. Dolphins (capable of abstract communication), primates (can use simple tools) and African Grey parrots (can categorise objects), for example, are intelligent in their own ways, but they all lack the most important aspect of human intelligence: its creativity that has resulted in the development of technology. Is evolution of creative intelligence inevitable? Once life appears on another world, what are the odds of its eventually evolving into creative intelligence? We already have a tag for such intelligence – ETI (extraterrestrial intelligence) – and an expectation of its creativity: a seat at the control panel of a spacecraft (UFO fans' belief), or at least at the control panel of an interstellar radio station (SETI enthusiasts' hope; SETI stands for 'search for extraterrestrial intelligence').

Ernst Mayr, the renowned evolutionary biologist who died at 100 in 2005, was not optimistic about ETI. He notes that out of probably more than a billion species of animals that have arisen on Earth, only one succeeded in producing the kind of intelligence to establish a civilisation. Even this civilisation did not develop the capability of interstellar communication until a few decades ago. He stresses that the assumption that any intelligent extraterrestrial life must have a technology and mode of thinking like us was unbelievably naive.

In an innovative essay published in 1985, Mayr illustrates the incredible improbability of intelligent life ever to have evolved, even on Earth, by representing the history of life on Earth on the scale of a calendar year:

Origin of Earth	1 January
Life (prokaryotes)	27 February
Eukaryotes	4 September
Chordates	17 November
Vertebrates	21 November

Mammals	12 December
Primates	26 December
Anthropoids	30 December at 1:00 am
Hominids	31 December at 10:00 am
Humans	31 December at 11:56:30 pm

This calendar shows that humans appeared 3½ minutes before the year's end; they occupy only 0.025 per cent of the total history of life on Earth.

The main thrust of Mayr's argument is as follows: What is most remarkable is that for about 2 billion years nothing very much spectacular happened as far as life on Earth was concerned. About 1.5 billion years ago a most remarkable event took place – the evolution of eukaryotes. (All organisms can be divided into prokaryotes (bacteria) and eukaryotes (all organisms except bacteria), the latter having a well-organised nucleus and chromosomes in each cell.) The most likely explanation for this event is the close association of two or more prokaryotes to create the first eukaryotes. After eukaryotes, an almost explosive innovative diversification took place. Within several hundred million years four new kingdoms evolved: the protists (one-celled animals and plants), fungi, plants and animals.

In none of these kingdoms, except that of the animals, Mayr says, was there even the beginning of any evolutionary trends towards intelligence. He illuminates the diversity of the animal kingdom and points out that, after more than 500 million years and numerous evolutionary lines, only primates showed true intelligence. It took another 25 million years and more evolutionary branches before one pathway eventually led to the rise of humans, less than one-third of a million years ago. There is no straight line from the origin of life to intelligent humans. 'The point I am making is the incredible improbability of genuine intelligence emerging,' he says. One inference of this point – that man is probably the most intelligent creature in the universe – should make the bishop's wife happy.

He points out another improbability: intelligence does not mean developing a technology capable of interstellar communication. Civilisations, as humans demonstrate, are fleeting moments in the history of an intelligent species. Even if two civilisations develop such a technology, it is necessary that they flourish simultaneously. He illustrates this improbability with a fable: 'Let me assume there is another high technology civilization in our galaxy. By some extraordinary instrumentation their inhabitants were able to discover the origin of the earth 4.5 billion years ago. At once they began to send signals to the earth and continued to do so for 4.5 billion years. Finally at the time of the birth of Christ they decided that they would terminate their program after another 1,900 years, if they had not received any answer by then. When they abandoned their program in 1900, they had proven to their own satisfaction that there was no other intelligent life in our galaxy.'

Mayr, however, does not want to 'deny categorically the possibility of extraterrestrial intelligence'. He just wants to claim that from an evolutionary biologist's point of view the probabilities are close to zero.

One day we will be the aliens

The general view is that humans are still evolving and, according to some experts, probably quite rapidly. So where are we heading? Ray Kurzweil, an American inventor and futurist, has an interesting take on this question. He says that in the not-too-distant future, humans will merge with technology and biological evolution will become obsolete.

This 'cutting-edge of evolution on our planet' will create something with smarter-than-human intelligence. He calls this new era 'the Singularity'.

The idea of machine intelligence has been around since 1950 when Alan Turing, the British mathematician famous for breaking the German navy's Enigma code during the Second World War, argued that it would one day be possible for computers to think like humans. But how would you find out whether your computer is as intelligent as you are? Turing suggested that if the response from the computer was indistinguishable from that of a human, the computer could be said to be intelligent. The Turing test has become the standard test for machine intelligence. A computer and a person are interrogated by text messages. If the interrogator cannot distinguish which answer came from the person and which from the computer, then the computer can be called intelligent. Are you and your computer ready for the test?

The day when you and I will flunk the test is not far away – at least, that's what Kurzweil tries to convince us in his book, *The Singularity is Near: When Humans Transcend Biology* (2005). 'As we get to the 2030s, the non-biological portion of our intelligence will predominate,' he says. 'By the 2040s it will be billions of times more capable than the biological part.'

Kurzweil divides the present and the future history of evolution into six epochs. 'The Singularity will begin with Epoch Five and will spread from Earth to the rest of the universe in Epoch Six,' he assures us. The six epochs are:

Epoch One: Physics and Chemistry. In this epoch, which started soon after the Big Bang, information was in atomic structure, especially in complicated, information-rich, three-dimensional structures of carbon atoms. The exquisite and delicate balance of the physical laws of the universe is precisely what is needed for the codification and evolution of information, resulting in increasing levels of order and complexity. Evolution, to Kurzweil, is a process of creating patterns of increasing order.

Epoch Two: Biology and DNA. In the second epoch, starting several billion years ago, DNA provided the precise digital mechanism for storing information and keeping it as a record of the evolutionary experiments of the epoch.

Epoch Three: Brains. The third epoch started when early animals developed the ability to recognise patterns, which still accounts for the vast majority of activities in our brains. Like earlier epochs, this epoch also provided a revolutionary or paradigm shift in the evolution of information: organisms now could detect information with their sensory organs and process and store that information in neural patterns in their brains.

Epoch Four: Technology. Mayr's calendar in the previous section shows the quickening pace of biological evolution. Similarly, according to Kurzweil, the pace of evolution of human technology is also quickening. This quickening is exponential (when we plot growth against time on a graph, a J-shaped curve shows exponential growth; a straight line shows linear growth); for example, it took only fourteen years from the PC to the World Wide Web.

Epoch Five: The Merger of Human Technology and Human Intelligence. When the Singularity arrives, the human-machines will 'overcome age-old problems and vastly amplify human creativity'. Does this mean no nukes, no wars and the human-machine race lives forever in harmony and peace? Kurzweil is not so sure: 'But the Singularity will also amplify the ability to act on our destructive inclinations, so its full story has not yet been written.'

Epoch Six: The Universe Wakes Up. In the last epoch in 'the evolution of patterns of information', our civilisation will infuse the rest of the universe with its creativity and intelligence (we suppose if it doesn't destroy itself in Epoch Five). Of course, Kurzweil realises that this infusion can happen only when human-machines can circumvent the limit that the speed of light imposes on the transfer of information. When we are able to supersede this limit, 'the "dumb" matter and mechanisms of the universe will be transformed into exquisitely sublime forms of intelligence'. 'This is the destiny of the universe,' he declares.

So why are we wasting our time searching for extraterrestrial intelligence when one day we will be that alien intelligence? When this question was posed to Kurzweil, his answer was succinct: 'It is worth trying to look for them anyway because the negative finding is just as significant as the positive one.' In his book he admits that it is likely (although not certain) that there are no other civilisations out there. 'In other words, we are in the lead ... in the lead in terms of the creation of complexity and order,' he says.

Back to Epoch Four. Kurzweil is dismissive of our version 1.0 biological bodies: they are frail, subject to myriad failure modes, and require cumbersome maintenance rituals. Each gene in this version is just one of 23,000 little software programs we have inherited that represent the design of our biology. 'It is not very often that we use software programs that are not upgraded and modified for several years, let alone thousands of years,' he says. It's time for an upgrade, and technology for this upgrade is already here. This technology, called RNA interference, is a tool for turning individual genes off. Help for the hardware upgrade will come from nanotechnology, which will enable us to create any physical product we want from inexpensive materials, using information processing.

'We will be able to go beyond the limits of biology, and replace your current "human body version 1.0" with radically upgraded version 2.0, providing radical life extension,' advises Kurzweil. The experience from software releases, says a wag, shows that nothing ever gets right until it hits version 3.0. So we must wait until Kurzweil's update kit for Human 3.0 hits the shelves.

And what would Epoch Five's Human 5.0 look like? Would this human-machine have a wireless broadband connection to universal intelligence and consciousness and look like Uncle Martin (*My Favourite Martian*, with two retractable antennae poking out of his head)?

If Earth is the zoo of an intergalactic civilisation, would they allow inferior Human 1.0 to become superior Human 5.0? That's another question – and another hypothesis.

Alien Big Brother is watching us all

Extraterrestrial intelligent life is widespread. Their reluctance to interact with us can be explained by the hypothesis that they have set aside our planet as part of a wilderness area or zoo.

This is the controversial and demoralising hypothesis proposed by John A. Ball in 1973. The hypothesis is based on three premises:

1. Whenever the conditions are such that life can exist and evolve, it will. Bell believes that the discovery of primitive life on Mars or anywhere else would probably solve this question.

2. There are many places where life can exist. Since Ball proposed this hypothesis, the discovery of extra-solar planets supports this premise.

3. We are unaware of 'them'.

Ball considers the third premise extremely significant. We are not aware of them because they are deliberately avoiding us and they have set aside our planet as a zoo or wilderness sanctuary. In a perfect zoo, animals would not interact with, and would be unaware of, their keepers. 'The hypothesis predicts that we shall never find them because they do not want to be found and they have the technological ability to ensure this,' he says.

He admits that the hypothesis is pessimistic and psychologically unpleasant: 'It would be more pleasant to believe that they want to talk with us, or that they would want to talk with us if they knew we are here.'

In a comment made in 1980 on the zoo hypothesis, Ball suggests that the civilisation that is number one exercises control and enforces the rules: 'They may keep us separate from our neighbours to prevent unfavourable or disastrous interaction.' Or perhaps they are waiting until the time is right to invite us to join the Galactic Club.

German astrophysicist Peter Ulmschneider also supports the zoo hypothesis. He said in 2003: 'While at first sight this idea appears quite extravagant, it makes considerable sense on closer inspection.' He suggests that our history would suffer drastic and fundamental changes – much more so than the history of the native Americans, annihilated when Columbus and Cortez arrived – if it came in contact with an alien advanced civilisation. 'It would be a catastrophe, a culture shock, and essentially an irresponsible act on behalf of the extraterrestrials,' he warns.

Ulmschneider, however, believes that a civilisation as capable of irresponsible acts as human civilisation has been in the past would not survive thousands, millions or billions of years without falling victim to the dangers of such behaviour. His conclusion: if highly advanced extraterrestrial civilisations exist, they have learned to act responsibly. This means 'under no circumstances will they disturb us or contact us, nor will they allow us to trace them by radio waves, through artefacts, or by direct contact'.

This rules out sending flying saucers to check on our well-being. Not a happy idea for UFO fans to entertain. Perhaps there is a sign somewhere at the perimeter of the solar system warning all trespassers: 'Wilderness Sanctuary: Don't Enter.' UFO fans can take solace in the knowledge that some cheeky alien teenagers do occasionally ignore this warning and fly their space buggies in our protected skies.

Ulmschneider is not so pessimistic about the future of the human race. He believes that one day we will have the knowledge to discover extraterrestrial intelligent societies in our own galaxy and other galaxies, but the possibility of ever directly interacting with them is very bleak. He is hopeful that, by that time, the basic idea behind the zoo hypothesis – that of acting responsibly – will become the guiding principle for our own behaviour. Moral: When we do escape from our zoo, we should become keepers of other zoos.

An extraordinarily rare Earth ...

Our planet is not the zoo of an extraterrestrial civilisation, but it's an extraordinarily rare planet. It has exactly the right location, moon, chemical composition and distance from the Sun to enable life to thrive. Its ideal location keeps it away from the rain of killer rocks, and thus safe from mass extinctions. Its large moon minimises changes in the planet's tilt, ensuring climate stability. It has enough carbon and other elements which help in the development of life and keep greenhouse conditions under control. Its right distance from the Sun ensures that water remains liquid, not vapour or ice.

These and many other unique features of Earth have convinced the University of Washington's Peter Ward, an authority on mass extinctions, and Donald Brownlee, an astronomer, to conclude that although microscopic life may be common on other worlds, complex life is rare. To them, even a flatworm is complex life. By this definition, the chances of existence of an extraterrestrial civilisation awash in technology are almost zero.

They argue in their book *Rare Earth: Why Complex Life is Uncommon in the Universe* (2000) that almost all environments in the universe are terrible for life, and Earth is probably the only Garden of Eden in the cosmos. They say that their rare Earth hypothesis, which has been called everything from 'simplistic' to 'brilliant' and 'courageous', is testable. They suggest two types of tests.

The first test is the search for microscopic life in other bodies in the solar system. This search can be done by specialised probes seeking life directly in samples. They suggest searching Mars, Jupiter's moons Europa and Ganymede, and Saturn's moon Titan for signs of alien microbes. The discovery of living micro-organisms or fossil evidence of micro-organisms will prove that life originates readily.

The second test involves the search for complex life forms. As there is no evidence of such life forms in the solar system, they suggest using powerful new telescopes to detect life on extra-solar planets.

As far as intelligent life is concerned, the two scientists support listening to alien radio signals. They advise against the more complex and costly exercise of beaming radio messages to nearby stars, and are doubtful whether the search for extraterrestrial intelligence is an effective use of resources. 'If the rare Earth hypothesis is correct, then it is clearly a futile effort,' they say.

... or a mediocre planet?

The 1971 'Communications with Extraterrestrial Intelligence' conference in Soviet Armenia generated more than one engrossing anecdote that lingered on in participants' memories (see page 54). At this conference, when a speaker droned on about his dubious theory that great scientific ideas are born during times when sunspots are active, Iosif Shklovsky whispered to Carl Sagan, who was sitting next to him, that this theory must have been conceived when sunspots were absent.

The witty Soviet astrophysicist died in 1985 at the age of 69, but his sharp wit can still be enjoyed in his riveting reminiscences, *Five Billion Vodka Bottles to the Moon: Tales of a Soviet Scientist* (1991). He was one of the first eminent astrophysicists to take a serious interest in the possibility of extraterrestrial intelligence. In 1962 he published a book in Russian, *Universe, Life, Intelligence*, which became an instant bestseller in the Soviet Union (the first printing of 50,000 copies was sold out in a few hours).

When Shklovsky asked Sagan to edit the English translation of this book, Sagan added so many notes of his own that they doubled the size of the book. As a result, his name appears as a co-author of *Intelligent Life in the Universe* (1966). In the book, Sagan's text is distinguished with triangles. This was a priceless boon to me, notes Shklovsky, otherwise vigilant Soviet censors could have made things tough for me. The book became the bible of the new SETI movement.

In this book, Shklovsky and Sagan ask: Is by 'some poignant and unfathomable joke' ours the only civilisation in the universe? Their answer is that 'our surroundings are more or less typical of any other region of the universe'. A high percentage of stars have planets, most of these planets are Earth-like, life has developed on most such planets, and intelligent life has evolved on a large number of these planets. There's nothing special about Earth. It's a mediocre planet.

From this assumption of mediocrity they conclude that 'all planets on which life has flourished for several billions of years have a high probability of the development of intelligence and technical civilization'. But they agree that this is at best a plausibility argument: 'we do not know the detailed factors involved in the development of intelligence and technical civilization.'

The idea of mediocrity was first proposed by German-American astronomer Sebastian von Hoerner in the early 1960s. He said that the Sun is an average star, Earth an average planet, and that humankind has an average intelligence. In *Cosmic Search* (January 1977), the first magazine devoted to SETI and astrobiology, he explains his assumption of us being average from one example: 'We have only one single case and that is us. And the question is, can we do statistics when the sample number N equals one? The answer is "yes", if you know the rules.

'If you have only one example to go by, with N equals one, then you do have an estimator for the average, and that is the one case you have. This would mean we should assume that we are average. On the other side with N equal to one, we do not have an estimator for the mean error. In plain English this means that the assumption that we are average has the highest probability of being right but we have not the slightest idea of how wrong it is.'

He concludes that if we generalise our own case, we should expect a very large number of extraterrestrial civilisations with whom we could talk.

Since they're not here, they do not exist

No, we will never get a chance to talk to an alien, if we accept American astrophysicist Michael H. Hart's bold argument for the absence of extraterrestrials on Earth: 'If there were intelligent beings elsewhere in our galaxy, then they would eventually have achieved space travel, and would have explored and colonised the galaxy, as we have explored and colonised Earth. However, they are not here; therefore they do not exist.' Before we say QED, let's ask the good scientist to demonstrate his proof.

In an imaginative paper first published in 1975, Hart starts his case by labelling the statement 'There are no intelligent beings from outer space on Earth now' as Fact A. He says that the argument presented in the previous paragraph is basically correct, but clearly incomplete. He puts all other explanations of Fact A into four groups:

1. Physical explanations: He dismisses the claim that aliens have never arrived on Earth because some biological or technological difficulty makes space travel impossible. For example, travelling at 10 per cent of the speed of light (about 1 billion kilometres per hour), a one-way trip to Sirius, one of the nearest stars, would take 88 years. He suggests that such large travel times may not be a problem to aliens if they have a lifespan of thousands of years. For aliens with a lifespan of 3,000 years, a voyage of 200 years is no different from the long sea voyages of our early explorers. Even if aliens do not have a long lifespan, it's not impossible to complete an interstellar journey in more than one generation 'if the spaceship is large and comfortable, and social structure and arrangements are planned carefully'.

2. *Sociological explanations:* He puts most proposed explanations of Fact A into this category. Three typical examples are: the *contemplation hypothesis* (mature civilisations are primarily concerned with artistic and spiritual pursuits and are not interested in visiting or colonising faraway planets); the *self-destruction hypothesis* (all advanced civilisations blow themselves up not long after they discover nuclear weapons); and the *zoo hypothesis.* 'No such hypothesis is sufficient to explain Fact A unless we can show that it will apply to every race in the galaxy, and every time,' he stresses.

3. *Temporal explanations:* They are not here simply because none have yet had the time to reach us. To judge the plausibility of this explanation, Hart assumes that we are indeed the first species in our galaxy to achieve interstellar travel. If we establish colonies on the 100 nearest stars, each of these colonies would establish their own colonies, and so on. He calculates that the time to traverse most of our galaxy would be about 650,000 years. 'We see that if there

were other civilizations in our galaxy, they would have had ample time to reach us,' he says.

4. *Perhaps they have come:* Hart discusses three versions of this hypothesis. The most common version is that visitors from space arrived here in the fairly recent past (within, say, the last 5,000 years), but did not settle here permanently. The weak spot of this version, according to Hart, is that it fails to explain why Earth was not visited earlier. The second version of the hypothesis is that they visited us a long time ago, say 50 million years ago. The third version, the UFO hypothesis, is that aliens have not only arrived on Earth, but are still here. Hart rejects this version: 'This version is not really an explanation of Fact A, but rather a denial of it.'

If his proof of Fact A is correct, Hart concludes, in the long run our descendants will probably occupy most of the habitable planets in the galaxy. These descendants might eventually encounter a few alien civilisations which never took any interest in interstellar travel, 'but their number should be small, and could well be zero'.

It's sad, but we're totally alone

American cosmologist Frank Tippler also believes that if extraterrestrials exist, they should already be here. 'Since they are obviously not, they do not exist,' he says. He does not deny the possibility that primitive life is widespread in the universe, but believes that the development of intelligence is vastly improbable. It has happened only once since the Big Bang. We are the first intelligence to evolve in the whole universe. 'We are totally alone,' he says.

Tippler's argument assumes that interstellar travel by self-replicating space probes is possible. If intelligent life were common, its emergence should have had a head start on planets around stars that are billions of years older than our Sun. At least one alien civilisation would have developed self-reproducing space probes and launched them into space. Travelling at 90 per cent of the speed of light – a technology not beyond the capabilities of an advanced civilisation, according to Tippler – it would take five years to reach the nearest star. If it takes 100 years to make a copy of itself, then the average speed at which all the probes would spread would be about $\frac{1}{25}$ of the speed of light. At this speed, Tippler estimates, the probes would spread throughout the galaxy within 10 million years. But we have no evidence of these probes on Earth. Their absence shows the absence of aliens. That's the logic of Tippler.

The idea of self-replicating machines was developed back in the 1950s by Hungarian-American mathematician and physicist John von Neumann. Self-replication would be possible if the hypothetical machine – von Neumann called it a 'universal constructor' – were provided with its own description as well as a means of copying and transmitting this description to the newly constructed machine. Such a machine has two components which are wholly distinct from one another – the machine and its description. The description serves two purposes: first, as a program that is executed during the construction of the new machine; second, as passive data that is duplicated and given to the new machine. Simply put, the machine has two parts: a computer and a constructor. The computer directs the constructor. The constructor, in turn, manufactures both another constructor and another computer. The original computer then copies its program to the new computer, which starts executing the program.

Von Neumann never applied his idea to space probes,

but many scientists including Tippler have done so since then. Tippler envisages his von Neumann probe as having human-level intelligence. It will be capable of using raw materials and energy resources available in distant star systems to replicate. On reaching a solar system, the probe will build several copies of itself. These copies will search for other solar systems, where the process will be repeated. 'Eventually, all the stars in the Galaxy would be reached by some descendant of the single original probe,' says Tippler. 'The Galaxy would be explored for the price of one von Neumann machine!'

His concept of a von Neumann probe goes a step further than simply making copies of itself to send to other stars. The probe would also be able to exploit mineral and energy resources of the star system to make technological artefacts, but he has no idea of the nature of these artefacts. Marcus Chown, a British science writer, agrees that this is not unreasonable: 'After all, the Romans would have had no idea that future civilisations would turn sand into computers, or bauxite into aeroplanes.'

These von Neumann probes could also be programmed to colonise space with replicas of humans. 'All the information needed to synthesise a human being is coded in the DNA of a fertilised human egg cell,' he says. 'This information could in principle be stored in the memory of a von Neumann machine, which could be instructed to synthesise an egg and place the "fertilised cell" in an artificial womb.' Thus our descendants one day will spread and colonise the entire universe. That's the dream of Tippler.

They're already in your backyard

Unlike Tippler, Paul Davies of the Australian Centre for Astrobiology at Macquarie University does not deny the existence of extraterrestrial intelligence. He speculates that an alien civilisation could use nanotechnology to build miniature space probes, not necessarily self-replicating probes, perhaps no bigger than your palm.

In 1959 the Nobel Prize-winning physicist Richard Feynman gave a classic talk, 'There's Plenty of Room at the Bottom', in which he said that 'the principles of physics, as far as I can see, do not speak against the possibility of manoeuvring things atom by atom' – or building molecular-sized machines. The advance of nanotechnology – the technology that deals with objects smaller than 100 nanometres (a nanometre is a millionth of a millimetre), especially manipulation of individual atoms and molecules – has fulfilled his prophecy.

'The tiny probes I'm talking about will be so inconspicuous that it's no surprise that we haven't come across one,' Davies says. 'It's not the sort of thing that you're going to trip over in your backyard. So if that is the way technology develops, namely, smaller, faster, cheaper, and if other civilisations have gone this route, then we could be surrounded by surveillance devices.'

In fact, he believes in the possibility of a probe resting on our own moon, having arrived at some time in our pre-history and remained to monitor our planet. American theoretical physicist Michio Kaku supports this idea. 'Personally, and this is a view shared by Paul Davies, a friend of mine, we believe that we have already been visited; and chances are they are on the Moon,' he says. The aliens are watching you. Let's find out how 'civilised' they are.

We lag behind in the civilisation stakes

The famous Russian astrophysicist Nikolai Kardashev was a doctoral student of Shklovsky at the Sternberg

Astronomical Institute in the early 1960s. In his reminiscences, Shklovsky is somewhat dismissive of Kardashev's and other scientists' optimism for extraterrestrial intelligence. He believes that 'the problem of extraterrestrial civilizations is in essence and effect *complex*'.

In spite of his teacher's misgivings, Kardashev was an earlier advocate of the idea that some alien civilisations may be billions of years ahead of us. In 1964, two years after obtaining his doctorate at the age of 30, he classified these advanced civilisations by a level of technological development that shows their use of energy resources for interstellar communications purposes. A Type I civilisation has enough energy resources to muster the equivalent to the entire present energy consumption of the planet Earth, a Type II civilisation the energy output of its own star, and a Type III the total energy output of its galaxy. We are not yet a Type I civilisation, according to this classification. Carl Sagan has classed our present civilisation as something like Type 0.7. At present we consume about one million billionth of the Sun's total energy received by Earth. Freeman Dyson has estimated that within 200 years or so we should attain Type I status, growing at a modest rate of 1 per cent per year.

Kaku has enhanced Kardashev's concept of classification of civilisations. According to him, a Type I civilisation can control the weather, earthquakes, tsunamis and volcanoes. Once they exhaust the power of a planet, they go to their nearby star. A Type II civilisation uses so much energy that 'their planet may glow like a Christmas tree ornament'. The only threat to their existence is a supernova explosion which may kill all life forms. A Type III is truly immortal, truly galactic in power and scope. It even has the power to change the evolution of a star before it explodes into a supernova.

On Kaku's scale, we qualify only for a Type 0 status. But he is optimistic, as he can see the seed of a Type I civilisation: 'We see the beginning of a planetary language (English), a planetary communication system (the Internet), a planetary economy (the forging of the European Union), and even the beginnings of a planetary culture (via mass media, TV, rock music, and Hollywood films).' If you speak English, work in Germany, read *Il Globo* on the Internet, listen to the Rolling Stones on your iPod and watch BBC TV sitcoms and Hollywood movies on your Nokia, you'll soon be allowed to wear a citizenship ring of the Shire of Greater Earth, a truly Type I civilisation. Citizenship has its privileges. The power to control the weather, earthquakes, tsunamis and volcanoes – on any Xbox in the universe.

It's all in the equation

American astronomer Frank Drake is a pioneering SETI researcher. In 1960 he became the first person in history to use a radio telescope to listen to ETs. In those early days of SETI, many scientists ridiculed the idea of extraterrestrial intelligent life, but to Drake the idea of other intelligent civilisations beyond Earth was a distinct possibility. He even placed a sign on his office door at the National Radio Astronomy Observatory in Green Bank, West Virginia: 'Is there intelligent life on Earth?' 'People would stop to read the sign, and then slowly smile at the way it framed the extraterrestrial question. Often as not, a head would poke inside my doorway with a wry comment such as, "There's a little green man downstairs who says he's looking for you", or "Just checking to see if there's any intelligent life in this room",' he writes in *Is Anyone Out There?* (1991), a book he co-authored with Dava Sobel.

In 1961 Drake invited a dozen scientists to the first-ever

SETI conference (now remembered as the Green Bank conference). While he was preparing for the conference he came up with an equation consisting of astronomical, environmental, biological and cultural parameters to estimate the number (N) of advanced civilisations that exist now in our galaxy and are capable of communicating across interstellar distances:

$$N = R_* \times f_p \times n_e \times f_l \times f_i \times f_c \times L$$
[astronomical] [environmental] [biological] [cultural]

where
R_* = number of stars born each year in our galaxy
f_p = fraction of these stars that have habitable planets
n_e = average number of planets or moons suitable for life around each star
f_l = fraction of worlds on which life actually appears
f_i = fraction of worlds on which life evolves to an intelligent form
f_c = fraction of worlds on which the intelligent life can communicate to other worlds
L = average lifetime of such technological civilisations

This equation is now known as the Drake equation. 'It amazes me to this day to see it displayed prominently in most textbooks on astronomy, often in a big, important-looking box. I've seen it printed in *The New York Times*,' writes Drake in his book. Since then the equation has become much more popular and ubiquitous. Even in the 21st century it continues to influence scientific thinking about intelligent life on other worlds.

But can it provide an accurate estimate? Some scientists are not so sure. James Gardner calls it 'speculative in the extreme'. Jill Tarter, a prominent SETI researcher (the character of Dr Ellie Arroway in Sagan's book and movie *Contact* was probably based on Tarter), says that it is 'really nothing more than a systematic way of quantifying our ignorance'. Nevertheless, if we assign values to the seven factors on the right-hand side of the equation, we can calculate N.

R_*: Of all the seven factors, R_* is the only one supported by observational evidence beyond the solar system. Participants at the Green Bank conference estimated that ten new stars would form each year. We know now that there are about 250 billion stars in our roughly 13-billion-year-old universe. Therefore, the average rate of star-birth is about twenty stars per year. But the figure of 250 billion stars includes all stars of all ages from 0 to 13 billion years, and does not include stars from the early days of the galaxy which have already died. So we can also say that the value of R_* is about 10.

f_p: The Green Bank group estimated this value to be about 0.5 (50 per cent), but ever-optimistic Sagan suggested a higher value of 1. We know now that not all stars can support habitable planets. Current estimates range from 0.2 to 0.9.

n_e: If we base our guess on the solar system, we can say that probably four worlds per solar system – Earth (absolutely), Mars (maybe in the past), Jupiter's moon Europa and Saturn's moon Enceladus (remote possibility) – are capable of sustaining life. Using our solar system as a guide, the Green Bank group estimated between 1 and 5 planets. Current estimates vary from 0.1 to 1.

f_l: Some scientists argue that this value is 1 because life is virtually inevitable on any habitable world; others give it a pessimistic value of one in a million, as they believe that the chances of life appearing on a planet or moon are extremely small. The Green Bank group also claimed that, given time, life will appear in a suitable environment. Charles Lineweaver and Tamara Davis say that the chance of life starting on an Earth-like planet is at least 1 in 3, that is, $f_l = 0.33$ (see page 50). Let's accept this value for our calculations.

f_i: The current estimates vary from a pessimistic 0.05 (based on the very long period of time it took intelligent life to evolve on Earth) to an optimistic 1 (once primitive life appears, it will surely evolve into intelligent form). The Green Bank group also concluded that wherever there was life in the universe it would show signs of intelligence, and gave f_i a value of 1.

f_c: This value can be taken as 1 on the simple reasoning that once you have intelligent life like ours, it's likely to develop capabilities to communicate at interstellar distances. The Green Bank group, however, arrived at an estimate of 0.2.

L: This is the most uncertain value in the Drake equation. We don't have any basis for a reliable estimate; even the only known technological civilisation has been communicating with radio waves for only about 100 years. Will our civilisation last a million years or will we destroy ourselves in the near future? Therefore, the value of L can range from an optimistic 10,000 to 100 million years (the Green Bank group's estimates) to a cautious 300 to 10,000 years (our estimates).

Now, by multiplying the seven terms we can arrive at a value for N:

Green Bank group's pessimistic estimate: $10 \times 0.5 \times 1 \times 1 \times 1 \times 0.2 \times 10,000 = \mathbf{10,000}$

Green Bank group's optimistic estimate: $10 \times 0.5 \times 5 \times 1 \times 1 \times 0.2 \times 100,000,000 = \mathbf{500,000,000}$

Our pessimistic estimate: $10 \times 0.2 \times 0.1 \times 0.33 \times 0.05 \times 1 \times 300 = \mathbf{1}$

Our optimistic estimate: $10 \times 0.9 \times 1 \times 0.33 \times 1 \times 1 \times 10,000 = \mathbf{30,000}$

These numbers show only what we already know: either we are alone or there are a large number of worlds with intelligent life that have the capability to communicate with us. The Drake equation is simply a mathematical way of saying 'we don't know'.

Fred Hoyle and Chandra Wickramasinghe quantify our ignorance with a different equation. They say that the number of technologies at any time is given by the formula:

number of technologies = technological life-span in years ÷ (life of a typical star ÷ number of habitable planets)

With 2 billion habitable planets and an average life of a typical star of 10 billion years, this gives:

number of technologies = technological life-span in years ÷ 5

If 300 years is taken as the typical technological life-span, we obtain only 60 technologies throughout the galaxy. If we take a much more optimistic estimate of 300,000 years, which is about the length of time *Homo sapiens* has been in existence: 'In this case the number of technologies in the galaxy turns out to be 60,000, with a grand total of 60 million for the entire visible universe,' they say.

Peter Ward and Donald Brownlee pour their *Rare Earth* cold water on this optimistic estimate by their own intimidating equation:

$$N = N^* \times f_\mathrm{p} \times f_\mathrm{pm} \times ne \times ng \times f_\mathrm{i} \times f_\mathrm{c} \times f_\mathrm{l}$$

where
N^* = stars in the Milky Way galaxy
f_p = fraction of stars with planets
f_pm = fraction of metal-rich planets
ne = planets in a star's habitable zone
ng = stars in a galactic habitable zone
f_i = fraction of habitable planets where life does arise
f_c = fraction of planets with life where complex animals arise
f_l = percentage of a lifetime of a planet that is marked by the presence of complex animals

They say that they have left out some of the more exotic aspects of Earth's history, such as plate tectonics and a large moon, from this equation. The point that Ward and Brownlee are making is that in any equation when a term approaches zero, so does the final result. If a planet does not have a right moon or its surface lacks conditions that can support water, then $N = 1$.

Hart has a better way to estimate N, without any equation: Assume some value for N and inquire about its effect on observables. In particular, he asks, if N is very large, then 'why aren't they here?' If there are a million civilisations in the Milky Way, he says, then at least one of them would have colonised our solar system by now, since the time to colonise the entire galaxy – with nuclear rockets, for example – is less than the age of the galaxy. 'Why are all the ETIs so shy?' wonders Kurzweil. According to American astronomer Thomas Kuiper, Fermi's paradox is a 'big gun in a small arsenal' of arguments in support of $N = 1$.

Let's roll a die to find a cosmic neighbour

If you find the idea of being alone in the universe gloomy, then turn to American statistician and author Amir D. Aczel for help. He has written a whole book on Drake's equation, *Probability 1: Why There Must Be Intelligent Life in the Universe* (1998), to show that there is intelligent life on at least one other planet.

First, Probability 101. Chevalier de Méré, a 17th-century high-living French nobleman and gambler, liked to bet even money that a 6 would come up in four rolls of a die. But when he started betting even money that a 6 would come up at least once in 24 rolls with two dice, his luck changed. He asked his friend, Blaise Pascal, one of the most brilliant mathematicians of his day, why he was having bad luck in his new game. Pascal wrote to fellow mathematician Pierre de Fermat about this problem, and their correspondence on the matter led to the birth of probability theory.

Chance is something that happens in an unpredictable way. Probability is the mathematical concept that deals with the chances of an event. We can find the probability of an event by dividing the number of ways in which the event can happen by the total number of possible outcomes. A die has six faces, numbered 1, 2, 3, 4, 5 and 6. The probability that any one of these numbers comes up is $\frac{1}{6}$.

What would be the probability of getting either a 3 or a

5? Because 3 and 5 cannot occur together, such an event is called a mutually exclusive event. In such events, probability is calculated by adding individual probabilities. The probability of getting either a 3 or a 5 is $\frac{1}{6} + \frac{1}{6} = \frac{1}{3}$.

When two dice are rolled separately, the second die does not take into account what the first die has done in order to decide what it will do. Such an event is called an independent event. In independent events, probability is calculated by multiplying independent probabilities. When two dice are rolled separately, the probability of getting a double 6 is $\frac{1}{6} \times \frac{1}{6} = \frac{1}{36}$. The probability of the complement of an event equals 1 minus the probability of the event. In other words, the probability of getting no double is $1 - \frac{1}{36} = \frac{35}{36}$. (The complement of an event is its opposite; for example, if in tossing a coin 'the coin shows heads' is the event, then the complement is 'the coin shows tails'.)

After this one-minute lesson in probability, we're ready to look at de Méré's problem.

Single die
Probability of a 6 = $\frac{1}{6}$
Probability of a number other than 6 = $\frac{5}{6}$
Probability of no 6 in four rolls = $\frac{5}{6} \times \frac{5}{6} \times \frac{5}{6} \times \frac{5}{6} = 0.48$
Probability of at least one 6 in four rolls = $1 - 0.48 = 0.52$ or
  52 per cent

De Méré's chances of winning his bet were 52 per cent. The odds, as punters say, were in his favour.

Two dice
Probability of a double 6 = $\frac{1}{36}$
Probability of no double 6s = $\frac{35}{36}$
Probability of no double 6s in 24 rolls = $(\frac{35}{36}) \times (\frac{35}{36}) \times \ldots$
  multiply 24 times = 0.51

Therefore, probability of at least one double 6 in 24 rolls =
  1 − 0.51 = 0.49 or 49 per cent

De Méré's chances of winning his bet were 49 per cent. The odds were against him. What are our odds of winning a cosmic neighbour?

In his book, Aczel takes the fraction of stars with planets, $f_p = \frac{1}{2}$. When he wrote the book, only nine extra-solar planets had been discovered, and only one was in the habitable zone. Similarly, Earth is in the habitable zone, the other eight planets possibly not, so he says we will use $\frac{1}{9}$ for this parameter. He considers DNA 'an extremely complex molecule with a very small chance of occurring on its own and that life is precarious because the universe is a dangerous place'. By this reasoning, he assumes the probability of life occurring on any single planet that is already within a star's habitable zone to be extremely remote, and assigns it an arbitrary value of one in a trillion. By multiplying $\frac{1}{2}$, $\frac{1}{9}$ and $^1/_{\text{trillion}}$, he arrives at a figure of 0.000,000,000,000,05, which is the probability of life around any one given star. He assumes that there are 300 billion stars in our galaxy, and there are 100 billion galaxies in the universe. He then puts all these values in the rule for combining the probabilities of independent events:

The probability of life on at least one other planet outside Earth with life on it = 1 − (0.999,999,999,999,95) × (0.999,999,999,999,95) × ... multiply 30,000 billion billion times = a number very close to 1.

'While we used the best scientific estimates, even lower values still lead to the same answer, a number close to 1,' Aczel says. 'The probability is virtual certainty.' We are not alone.

Aliens have evolved into machines

If we ever meet our sole neighbours on a world beyond
Earth populated by Aczel's statistical wizardry, they're
likely to be machines, not creatures made of carbon and
water. That's if we believe Steven J. Dick, an American
astronomer and historian of science.

Olaf Stapledon, the British author and philosopher who
died in 1950, is remembered primarily for his fictitious 'his-
tories of the future'. His science-fiction novel *Last and First
Men* (1930) traces the history of humanity from the First
Man (that's us) through the Eighteenth Man or the Last
Man (living on Neptune 2 billion years hence).

Dick says that if biology and culture exist beyond Earth,
they will evolve; and to understand this evolution we must
think not only in astronomical timescales, but in what he
calls Stapledonian timescales, which take into account the
evolution of biology and culture. If we apply Stapledonian
thinking to biological and cultural evolution, it will make
extraterrestrial intelligence far different from us.

He believes that we inhabit a post-biological universe, a
universe in which the majority of life has evolved beyond
flesh and blood intelligence. His post-biological universe is
based on three premises:

1. The maximum age (*A*) of extraterrestrial intelligence is
   several billion years.

2. *L*, the term in Drake's equation, is greater than 100 years
   and probably much larger.

3. In the long term, cultural evolution overtakes biological
   evolution.

If we assume that we can survive nuclear world wars, mass extinctions and other natural catastrophes such as asteroid impacts, 'L could conceivably approach A, which is billions of years'. Dick admits that we cannot know with certainty 'how long does L have to be before we reach a post-biological outcome?' By extrapolating current trends in artificial intelligence, biotechnology and nanotechnology, he estimates that we will certainly make the transition to a post-biological universe if L is greater than 1,000 years. He accepts that this result is based on applying the insights of Kurzweil, Tippler and others to the entire universe, using Stapledonian thinking.

What would this post-biological universe be like? A *Westworld* (1973) filled with gunslinging Yul Brynner robots? We hope not. Dick lists three characteristics of post-biological 'humans'. First, they will be immortal; they will have the capability to repair and update. Second, their capacity for great good or evil. Third, as nothing in the universe remains static, the post-biologicals will also evolve into creatures that 'might have all the characteristics we ascribe to God: omniscient, omnipotent and perhaps [with] the capability of communication through messenger probes'.

Welcome to a Futureworld of 'self-improving thinking machines'. Where are all those cute ETs? Steven J. Dick, please step aside. Let Steven Spielberg write the scenario for the post-biological universe. Meanwhile, we earthlings shall wait for whispers from space.

CHAPTER 6

Whispers

The mystery of the 'Wow' signal

It came from outer space and lasted only 72 seconds. On the computer printout it simply appeared as 6EQUJ5 – a code that revealed it was a strong, intermittent radio signal confined to a narrow band of frequency. It was so unusual that it caused an excited astronomer to scrawl 'Wow!' in the margin of the printout, a label that is inextricably linked to it.

What was it? A message from an alien intelligence? A momentary hiccup from a cosmic event, or a polluting burp from a terrestrial transmission? Three decades on, no one knows what really created the signal, and the debate continues.

The Big Ear radio telescope at the then Ohio State University Radio Observatory had been involved in the search for extraterrestrial intelligence since 1973. On the night of 15 August 1977 its 79-metre dish was tuned to 1,420 megahertz, the frequency of hydrogen atoms.

Interstellar space is filled with hydrogen atoms, at a density of about one atom per cubic centimetre. The individual atoms chirp at a frequency of 1,420 megahertz, like a miniature radio station. The chorus of these countless atoms can easily be heard by any radio telescope. As hydrogen is the most common element in the universe, many people believe that the aliens might choose this frequency to broadcast their presence to us. Even if the aliens do not use this frequency, they are likely to send a narrow-band signal, which has a precise frequency and a longer range. A broad-band signal, on the other hand, is like a noise that covers a wide range of frequencies. Stars and many other cosmic objects emit broad-band noise. Radio stations broadcast narrow-band signals, but the hiss between stations is broad-band noise. The terrestrial intelligence advises that an alien intelligence would not want their message to be mistaken as ordinary star noise.

For this reason, the Big Ear had been converted from measuring the location and strength of wide-band signals to narrow-band signals. In a radio telescope, signals received from a large area are focused by the dish to a receiver that amplifies the fluctuating voltages of the signals. You can't just plug your headphones into the receiver and listen to the chatter, natural or alien-made. Even if you could, all you would hear is a hissing sound. The Big Ear used a computer to record the signals. The floor-standing desk-size IBM 1130 used by the observatory was equipped with 16 kilobytes of RAM and a megabyte of storage on a magnetic disk. This stone-age storage capacity meant that the daily observations had to be printed out and examined by volunteers, as the observatory did not have any funding for the programme.

Jerry Ehman, a professor at Ohio State University, took the responsibility for this task. A few days after 15 August, he began his routine review of the printouts, not expecting to find anything unusual. As he worked his way through the reams of paper for the night of 15 August, he was astonished to see the string of numbers and characters 6EQUJ5 on the printout. It represented a burst of radio waves, like a thunderclap in the middle of a piece of quiet music. Ehman immediately recognised it as a narrow-band signal confined to a frequency around 1,420 megahertz. Without thinking, he wrote 'Wow!' and circled the string 6EQUJ5 on the printout. 'It was the most significant thing we had seen,' recalls Ehman. For a month following the discovery, he and his colleagues looked for the signal again at least 50 times, but found nothing.

Each character of the alphanumeric code used by the observatory's computer represented the strength of a received signal over twelve seconds. The code 6EQUJ5 revealed that the 'Wow' signal rose and fell over the course of 72 seconds:

6	6
E	14
Q	26
U	30
J	19
5	5

(The numerals in the right-hand column show approximate signal-to-noise ratios; that is, how many times the signal is stronger than the background noise. For example, the strongest signal received, U, was about 30 times stronger than the background noise. Most of the background noise is generated within the receiver itself; some noise comes from buildings, trees, grass and other surroundings, and the celestial sky.)

The fact that the signal rose and fell over the course of 72 seconds is intriguing. The Big Ear was a fixed telescope (it

Figure 5. A section of the computer printout showing Jerry Ehman's 'Wow!' and the circling of 6EQUJ5 (Image: Jerry Ehman, North American Astrophysical Observatory, the continuing organisation of the Big Ear Radio Observatory)

was demolished in 1998), and the Earth's daily spin allowed it to pick up cosmic signals from a tiny angular section of the heavens. The string 6EQUJ5 shows that as the radio source passed by, its intensity rose as the Earth's spin brought it within the telescope's range, reached a peak in the centre, and then faded away. For the Big Ear, this rise and fall should last 72 seconds, and that's what happened. If the signal were from a terrestrial source, it would suddenly flood the telescope and then switch off after some time.

Under certain conditions, a terrestrial signal reflected off a piece of space debris could appear as if it were coming from a point source like the Wow signal. 'I now place a low probability on this alternative,' says Ehman. A terrestrial signal also scores low probability for another reason: radio transmissions in the frequency band around 1,420 megahertz are prohibited by international agreement.

Did the Wow signal include a message? The signal was unmodulated. Unlike modulated AM or FM radio signals (in which characteristics of the signals are changed or modulated to include data), it was a blast of radio noise. However, it is still possible that there could be modulation at frequencies that were not within the range of the observatory's detection equipment. 'We would not have seen that modulation, and hence we could say that modulation is within the realm of possibility,' remarks Ehman.

Ehman and his colleagues also looked at other possible sources of the signal:

Planets: None of the planets was close to the signal source position. Planets do generate some radio waves, but they have much broader frequencies.

Asteroids: Asteroids are essentially small planets. None of the larger asteroids was in the vicinity.

Satellites: An investigation of the orbits of all known satellites showed that none was within the telescope's range. Satellites do not use the protected frequency band of 1,420 megahertz.

Aircraft: Aircraft also do not transmit in the protected band. An aircraft will also show a significant motion with respect to the stars, which would cause the pattern of signal to be very different from that expected from a point source.

Spacecraft: Spacecraft are also prohibited from transmitting in the protected band; none was within the telescope's range.

Interstellar scintillation: We see the stars twinkling because their light is scattered unevenly by the Earth's atmosphere. Similarly, when light and radio waves travel through interstellar space, they also 'twinkle'. It is possible that a radio signal could become stronger as it passes through interstellar space. If it did occur to the Wow signal, it still points to a signal coming from a source many light years away from us. 'This gives more support for the hypothesis of a signal of an extraterrestrial origin,' says Ehman.

Gravitational lensing: Gravity can bend light and radio waves. Did the gravity of an intervening star focus a much weaker signal into the stronger Wow signal? Ehman rules out this explanation on the ground that gravitational lensing usually lasts many days or months depending on the motion of the source.

So the evidence stacks up in favour of an alien source. Not quite. The Big Ear used two funnel-shaped metal structures called horns, situated side by side, to collect radio waves focused by the dish. The path of the radio waves collected by a horn is called a beam. As Earth rotates, any cosmic radio source would be seen first in beam one (for 72 seconds) and then – about three minutes later – in beam two (also 72 seconds). Ehman's computer printout showed it only on one beam, instead of the two beams expected. The computer was not programmed to identify whether the signal came from the first or the second beam. 'Suspicious and disheartening,' remarks Seth Shostak of SETI Institute, a California-based non-profit organisation that conducts research on life in the universe.

Perhaps the aliens turned off their transmitter after three minutes and went on holiday. Was the signal a wish-you-were-here postcard? No one knows. Unlike alien astronomers, terrestrial astronomers are workaholics. From observatories around the world they have performed more than 100 searches of the same region of the sky. No report yet of an astronomer running naked through the street, shouting 'Wow! Wow!'

Robert Gray, an independent SETI researcher from Chicago, has been trying to solve the riddle of the Wow signal for more than a decade. In 2001, using a radio telescope 100 times more sensitive than the Big Ear, he looked for a radio source – artificial or natural – at the position that Wow came from, but found nothing. In a recent survey, with Simon Ellingsen of the University of Tasmania, he looked at the possibility of a periodic source to explain the single-beam detection of the Wow signal. Pulsars, stars that emit rapid pulses of radio waves, are familiar examples of cosmic periodic sources of radio waves. Using a 26-metre radio telescope, Gray and Ellingsen made six fourteen-hour observations of the Wow spot in the sky. Their conclusion: 'No signals resembling the Ohio State Wow were detected.'

Calculations show that the Wow signal came from the direction of the constellation of Sagittarius; the nearest star in the constellation is 220 light years away. If someone were sending a message from such a faraway star, they would need a large 300-metre dish and a gigantic 2,000-megahertz transmitter. 'That's daunting, but not altogether impossible,' assures Paul Shuch, an American SETI enthusiast. Even if we believe Shuch, the message might not have been meant for us. It might have been a message from an alien civilisation to another alien civilisation. Did the Big Ear inadvertently eavesdrop on an interstellar phone conversation?

Ehman believes that even if the Wow signal were an alien signal, they would do it far more than once: 'We should have seen it again when we looked for it fifty times.' Shostak agrees: 'So until and unless the cosmic beep measured in Ohio is found again, the Wow signal will remain What signal.'

A tiny blue spark heralds the radio age

James Clerk Maxwell, whose scientific achievement is comparable with that of Newton or Einstein, showed so little promise at school that his classmates nicknamed him 'Dafty'. But when his father took him to a science lecture he soon became interested. A few years later he sent his first scientific paper to the Royal Society of Edinburgh. The paper presented a novel technique for drawing a perfect oval, and was so well done that no one could believe it was written by a fourteen-year-old boy.

In 1864, when he was 33 years old, Maxwell read to the Royal Society of London his most important paper, showing a connection between electricity and magnetism. The paper also included his four equations – now famous as Maxwell's equations – that express mathematically the way in which electricity and magnetism behave. Though the equations predicted the existence of electromagnetic waves, which travel at the speed of light and consist of electric and magnetic fields vibrating in harmony in directions at right angles to each other, Maxwell never proved their existence.

In 1886, seven years after Maxwell's death, Heinrich Hertz, a newly married young professor at the Technische Hochschule in Karlsruhe, Germany, set out to find the elusive 'wireless waves', as they were then known. He modified an induction coil to generate sparks across a gap between two brass balls. In those days it was a common set-up for demonstrating electric discharge. A few metres away from the induction coil he placed a loop of wire connected to two brass balls separated by a tiny gap. When he passed the discharge through the induction coil, he was startled to see a tiny spark in the loop of wire a short distance away. His wife was with him to witness the first radio transmission in history: one wave emitted at one point had been received at another. No music, no talkback shows, just a tiny blue spark and an entry in Hertz's diary for 1 November 1886: 'Vertical electric vibrations, in wires stretched in straight lines, discovered; wavelength, 3 metres.'

If Maxwell, who loved reading and writing poems, were alive, perhaps he would have rewritten his poem, 'Valentine from a Telegraph Clerk ♂ to a Telegraph Clerk ♀', for the new age of radio. An excerpt:

> 'O tell me, when along the line
> From my full heart the message flows,
> What currents are induced in thine?
> One click from thee will end my woes'.
> Through many an Ohm the Weber flew,
> And clicked the answer back to me,
> 'I am thy Farad, staunch and true,
> Charged to a Volt with love for thee'.

(In Maxwell's time, the term 'Weber' was used for Ampere and 'Farad' for Coulomb.)

The tiny blue spark proved that electromagnetic waves do exist. A year later, Hertz was able to measure the speed of these waves and to show that the speed was the same as that of light. His further experiments showed that electromagnetic waves could be refracted, reflected and polarised in the manner of light.

Hertz never realised the importance of his discovery.

When he demonstrated his experiment to his students, someone asked what the discovery could be used for. 'Nothing, I guess,' replied Hertz. It was left to the Italian physicist Guglielmo Marconi and others to develop technology for the practical use of Hertzian waves.

We are now familiar with all types of electromagnetic waves or radiation that make up the complete electromagnetic spectrum. These waves are streams of photons travelling in a wave-like motion with the speed of light. Photons are bundles of energy. They have no mass and their energy increases with their frequency. The only difference between various types of electromagnetic waves is the energy of the photons. In order of increasing energies, electromagnetic waves are called radio waves, microwaves, infrared, visible light, ultraviolet, X-rays and gamma rays. This order also shows increasing frequencies or decreasing wavelengths. Wavelength, the distance between two successive crests or troughs of a wave, is measured in metres. Frequency, the number of crests – or complete cycles – that pass a given point in one second, is measured in hertz (Hz), a unit named in Hertz's honour.

Our noisy cosmic neighbours

While Hertz was looking at ways of sending 'wireless waves' through air, British physicist Oliver Lodge was experimenting with transmitting these waves along electric wires. Hertz's discovery overshadowed his work. Nevertheless, the feisty physicist continued his research on electromagnetic waves and in 1894 suggested that if heat, light and radio waves were different manifestations of electromagnetic waves, radio waves might also be coming from the Sun to Earth. In 1897 he attempted some experiments to pick up radio waves from the Sun by blocking

heat and light with an opaque substance. He admits the
failure of his experiments in his book, *Signalling across Space
without Wires* (1897): 'There were evidently too many terrestrial sources of disturbances in a city like Liverpool to make
the experiment feasible.' Lodge's failure to communicate
with the Sun (curiously, he spent his later life experimenting with the possibility of communicating with the dead)
meant that the world had to wait another 36 years to hear
the news of the discovery of extraterrestrial radio waves
and the birth of radio astronomy.

The news first appeared in the *New York Times* of 5 May
1933, and filled an entire front-page column. The report,
headlined NEW RADIO WAVES TRACED TO THE CENTRE OF THE
GALAXY, denied that they were signals from extraterrestrials:
'There is no indication of any kind ... that these galactic
radio waves constitute some kind of interstellar signalling,
or that they are the result of some form of intelligence
striving for intra-galactic communication.'

On 27 April 1933, Karl Jansky, a 37-year-old engineer
from the Bell Telephone Laboratories, read a paper entitled
'Electrical Disturbances of Extraterrestrial Origin' to a small
audience at a meeting of the International Scientific Union
in Washington, DC. The paper, which received a lukewarm
reception, is now viewed as the beginning of radio astronomy. A few days later, Bell Labs' spin doctors decided to
send out a press release. Media organisations all over the
world picked up the news and Jansky became an instant
celebrity. On 15 May, Jansky appeared on an NBC radio
programme. Listeners throughout the United States heard
the hiss of the stars live by a telephone hook-up to the
receiver of Jansky's primitive radio telescope at Holmdel in
New Jersey. The next day, the *Times* headlined: RADIO WAVES
HEARD FROM REMOTE SPACE ... AFTER TRAVELLING 30,000 LIGHT-

YEARS. The report described the hiss as 'sounding like steam escaping from a radiator'.

When Jansky joined the Bell Labs in 1928 as a radio engineer, the recently opened New York to London radio telephone service was plagued with intermittent noisy crackling of static interference caused by various sources such as electrical equipment and thunderstorms. Bell Labs' research engineers were interested in reducing noise levels in telephone conversations. Jansky's job was to record the intensity of this interference with a radio receiver connected to an antenna – a long array of metal pipes – mounted on four Ford Model T wheels. An electric motor moved the wheels on a 3-metre sprocket to point the antenna to any part of the sky. During one of his experiments, Jansky heard a steady hissing sound at a frequency of 20.5 megahertz (or a wavelength of 14.6 metres), which was very different from the intermittent crackling of the static. He had the insight to realise that the sound was not of terrestrial origin. He also ruled out the Sun as the possible source. He soon discovered that the signals became louder whenever the antenna pointed towards the constellation of Sagittarius. The young engineer had accidentally opened a new window to the universe. But it took scientists a long time to understand the significance of Jansky's discovery and look through this new window. Even Jansky did not pursue this new field further.

No story of radio astronomy is complete without a mention of the pioneering work of Grote Reber. He was a 22-year-old student at the Illinois Institute of Technology in Chicago when he heard about Jansky's discovery. He was so stimulated by it that he applied to almost all major research centres for astronomy in America to continue the research. No one showed any interest. Radio astronomy was a strange beast to them. In 1937, Reber decided to build a visionary 9-metre bowl-shaped antenna in his backyard. Using this first true radio telescope, for nearly a decade he conducted an extensive survey of the sky and produced the first maps of radio sources in the galaxy. The world was now ready to hear from our noisy cosmic neighbours.

Listening to the 'songs' of the cosmos

Because the wavelength of radio waves is much larger than the wavelength of light – some are longer than a kilometre – radio telescopes are much larger than optical telescopes. Radio telescopes are metal dishes that focus radio waves to a point. The spherical dish of the world's largest telescope, at Arecibo in Puerto Rico, is 305 metres in diameter, 51 metres deep and covers an area equal to seven football fields. Suspended 137 metres above the dish is a platform that houses a set of movable antennae and a highly sensitive radio receiver. To make radio images sharper, radio astronomers also combine many smaller receiving dishes into an array. The Very Large Array Radio Telescope in New Mexico consists of 27 antennae arranged in a huge 'Y' shape up to 36 kilometres across. This arrangement works as a radio telescope with a virtual dish 36 kilometres in diameter.

Most astronomical objects – planets, asteroids, comets, stars, giant clouds of gas and dust, black holes and galaxies – emit radio waves. According to the Science@NASA website, if we had radio antennae instead of ears, we would hear a remarkable symphony of noises coming from our own planet. These noises are around us all the time. Scientists call them 'tweaks' (they sound like a quick musical ricochet), 'whistlers' (sound like a musical descending tone) and 'sferics' (short for 'atmospherics', sound like twigs snapping or bacon frying). These terrestrial radio waves can be converted into audible sound by a very low frequency (VLF) radio receiver.

Jupiter is another source of exotic sounds. As one would expect from the biggest planet in the solar system, these sounds are much louder than terrestrial sounds. Two types of Jovian 'songs' can be heard by a radio receiver with a special antenna in the shortwave bands between 15 and 40 megahertz: 'L-burst' (sounds like ocean waves catching on a distant beach) and 'S-burst' (a staccato of rapid popping sounds with a beat that reminds one of woodpeckers).

The Sun is the most intense source of radio waves in the solar system. Artificial radio signals from Earth come next. Whenever there are sunspots or solar flares, the intensity of solar radio waves increases. These waves can be heard directly, particularly at dawn, on a shortwave radio receiver at frequencies close to 20 megahertz.

Outside the solar system, the sources of radio waves include clouds of gas and dust, stars of various kinds such as pulsars (rapidly rotating neutron stars), quasars (short for 'quasi-stellar radio sources', galaxies during early stages of their life), black holes (a region of space-time from which nothing can escape, not even light) and galaxies (usually elliptical galaxies). The source of Jansky's radio hiss was most likely a supermassive black hole at the centre of our galaxy.

The idea that launched a thousand searches

To extraterrestrial 'civilizations with scientific interests and with technical possibilities much greater than those now available to us', ... 'our Sun must appear as a likely site for the evolution of new society. It is highly probable that for a long time they have been expecting the development of science near the Sun. We shall assume that long ago they established a channel of communication that would one day become known to us, and they look forward patiently to answering signals from the Sun which would make known to them that a new society has entered the community of intelligence.'

This is an extract from the first two paragraphs of a seminal paper published in 1959 by Philip Morrison and his Cornell University colleague Giuseppe Cocconi. The paper was the first to suggest the idea of searching for extraterrestrial intelligence by radio astronomy.

'What sort of a channel would it be?' they ask. Their answer: electromagnetic waves. They say that 1,420 megahertz is the ideal frequency: 'It is reasonable to expect that sensitive receivers for this frequency will be made at an early stage of the development of radio-astronomy. That would be the expectation of the operators of the assumed source, and the present state of terrestrial instruments indeed justifies the expectation.'

They end their paper with the exhortation that these speculations should not be confined to the domain of science fiction: 'Few will deny the profound importance, practical and philosophical, which the detection of interstellar communications would have. We therefore feel that a discriminating search for signals deserves a considerable effort. The probability of success is difficult to estimate; but if we never search, the chance of success is zero.'

A year later, in 1960, 29-year-old Frank Drake made the first real attempt to eavesdrop on radio broadcasts from extraterrestrials. He aimed a 26-metre radio telescope at the stars Tau Ceti and Epsilon Eridani, some eleven light years away. The telescope's receiver was tuned to a frequency of 1,420 megahertz. On the very first day of the project, recalls Drake, 'Wham! A burst of noise shot out of the loudspeaker, the chart recorder started banging off the scale, and we were all jumping at once, wild with excitement.' The strong, clear signal was precisely 'what you would expect from an extraterrestrial intelligence trying to attract attention – as though they'd just been waiting for us to tune in'. The signal was probably caused by a secret US military experiment. The erroneous idea that Frank's team made a secret discovery of the project and its 'association' with the US military is what conspiracy theories are made of. The signal has earned a place in the folklore of UFO fans.

The project – dubbed Project Ozma after the queen in *The Wizard of Oz* (Oz was a place 'very far away, difficult to reach, and populated by strange and exotic beings') – lasted only two months. Except for that short signal, all they heard from the loudspeaker was static, and the chart recorded nothing but formless wiggles. After nearly half a century, numerous other SETI projects have not done any better.

Waiting at the waterhole to say hello to ET in 2025

Radio waves are not absorbed by interstellar gas and dust, which block visible light. They require only modest power and inexpensive equipment to transmit, and are easy to detect. These two qualities make them the method of choice for any interstellar civilisation wanting to make itself known to other interstellar civilisations.

Within the radio spectrum, SETI researchers prefer to look for signals in the so-called 'waterholes'. The first waterhole is between 1,420 and 1,638 megahertz (wavelengths 21 and 18 centimetres). The first frequency is associated with hydrogen (H) and the second with hydroxyl radical (OH). When H and OH combine, they form water (H_2O) – hence the name waterhole. Extraterrestrial life forms based on water, like us, are likely to use these frequencies. The second waterhole is around 22,000 megahertz (wavelength 14 millimetres), the frequency associated with water. The third waterhole is the cosmic microwave background – frequency 150,000 megahertz or wavelength 2 millimetres – which is explained in the next section.

When searching for intelligent radio signals, SETI researchers face two choices: should they make 'targeted' searches or 'wide-sky surveys'? In a targeted search, researchers examine individual nearby stars with high-sensitivity radio telescopes. In a wide-sky survey, a larger swathe of the sky is scanned with lower sensitivity. Seth Shostak does not consider the two strategies as complementary. He would like to abandon star-by-star targeting in favour of scanning the areas of sky that have a greater number of stars, even if most of them are very far away. 'Unless ETs truly infest the stars like flies (very unlikely),' he says, 'the first signals we can detect will come from very rare, very powerful transmitters very far away.'

If aliens are not transmitting signals deliberately, we may have to eavesdrop. For eavesdropping, TV signals are ideal. The Earth's atmosphere absorbs most radio signals, but higher-frequency TV signals travel past the horizon into space and can then continue for distances of hundreds of light years. It is possible that right now someone 54 or so light years away is watching the coronation of Queen Elizabeth II, broadcast live on BBC on 2 June 1953. Similarly, if we ever receive a leaked extraterrestrial TV signal, it would have been transmitted decades ago, even centuries or millennia ago.

Intelligent aliens looking for TV signals as signs of life in our solar system must hurry up, as 'Earth is going to disappear very soon'. That's the advice from Frank Drake. He's worried that within 50 years or so, cable and direct-broadcast satellites will replace terrestrial TV transmission. Cables do not leak any signals into space, and the power of satellite broadcast is very low, about 20 watts per channel, all efficiently directed straight down towards Earth's surface. For humans, TV signals 'are the strongest signs of our existence', Drake says. Most SETI researchers also focus on eavesdropping alien signals produced for a planet's domestic use, such as our own TV, FM radio and radar signals, but inadvertently leaked out into space. Such signals are weak and difficult to spot. A beacon – a coded repetitive signal beamed to attract attention – from an alien civilisation is perhaps SETI researchers' best hope.

George W. Swenson, Jr., an American electrical engineer, is not optimistic about radio searches. He has calculated that the power required for sending a signal in all directions to a distance of 100 light years is 5.8 million billion watts, which is 7,000 times the total electricity-generating capacity of the United States. There are only about 1,000 stars within 100 light years, and the kind of systems that would be needed to mount a realistic project to beam a signal to a large number of them are probably beyond the resources of our society. Even if contact could somehow be made, the time delay before a response to a message could be received might very well stretch into many centuries. 'This is clearly a project for many generations in succession,' he warns. 'In all likelihood, it will require an enduring organization based on immutable dogma – like one of the world's major religions.'

Shostak has an immutable belief in intelligent aliens and their society's capacity to generate radio signals. He also has faith in terrestrial technology. He believes that advances in computer processing power and radio tele-scope technology will ensure that we detect alien signals within the next two decades. Shostak has called upon Drake's equation and Moore's law for his prediction. Using Drake's equation, he has arrived at an estimate of between 10,000 and 1 million alien radio transmitters broadcasting in our galaxy. He believes that Moore's law – that computer processing power doubles every eighteen months – will hold true for the next ten years (as it has done for the past 40 years). For the following ten years, he takes a conserva-tive estimate of a doubling of computer processing power every 36 months. This increased processing power will allow the analysing of radio signals from all stars between 200 and 1,000 light years away. We have about two decades to think about what we are going to say in our own message, and then a wait for at least two centuries to hear the reply from our friendly aliens. But there's already a message for us in the sky, if we can read it.

A hidden message in the sky

Our universe began when an unimaginably dense and unimaginably hot speck of matter exploded spontaneously. The newly born universe was so hot that electrons and nuclei could not combine to form matter. The free-moving electrons scattered the photons, making the universe opaque. After 380,000 years, the universe cooled to about 4,500 degrees and electrons and nuclei could combine to form the first hydrogen atoms. The universe now became transparent and photons were free to escape as high-energy gamma rays. As the universe continued cooling and expanding, the wavelength of the radiation stretched. It changed from short wavelength gamma rays to longer wavelength X-rays, ultraviolet rays, visible light, and after

13.7 billion years into microwaves. The remnant radiation, usually referred to as the cosmic microwave background, has a frequency of 150,000 megahertz (wavelength 2 millimetres); and its temperature is very cold, about –270 degrees Celsius (3 degrees above absolute zero). This leftover warmth from the primeval fireball fills the universe today. If we could see it, the entire sky would glow with astonishingly uniform brightness in all directions.

To American theoretical physicists Stephen Hsu and Anthony Zee, the cosmic microwave background is like a giant billboard in the sky, visible to every technologically advanced civilisation in the universe: 'Anybody can see it, regardless of whether you have five heads and three eyes, even if your biology is not based on carbon.' They believe that this billboard is the ideal place if the universe's creator wanted to leave a single universal message for its future inhabitants.

Cosmic microwave background has tiny temperature variations that could serve as a string of zeros and ones for a binary message. In 2001 NASA launched its WMAP satellite to measure the temperature variations of the cosmic microwave background, and it has since produced maps of the microwave sky. These maps will help scientists answer fundamental questions about the origin and fate of the universe. The graphs of WMAP measurements of temperature differences between pairs of points on the sky plotted against their angular separations create curves with series of peaks and troughs. These peaks and troughs so far have not revealed any secret code. It may be twenty or thirty years before satellites are sensitive enough to collect accurate data to determine whether Hsu and Zee's idea has any merit or if it's just plain wacky.

The physicists estimate that the maximum amount of

information that could be written in a microwave message is about 100,000 bits. The first few bits must specify the key; otherwise no one will ever be able to understand it, even if they stumble upon it. What do Hsu and Zee expect to find in the message? The Holy Grail of physics, of course, the elusive theory of everything. They believe that the equations describing the fundamental laws of physics will make sense to all civilised life forms in the universe and are compact enough to fit into a small chunk of bits.

Hsu and Zee urge that when more accurate cosmic microwave background data becomes available, it should be analysed for possible patterns. To them, looking for a message in the sky is more fun than searching for messages from aliens.

CHAPTER 7

Messages

Incongruent triangles

It has been said of Carl Friedrich Gauss (the 'Prince of Mathematicians' who can be ranked only with Archimedes and Newton, 'and it is not for ordinary mortals to attempt to range them in order of merit', warns E. T. Bell) that almost everything which the mathematics of the 19th century has brought forth in the way of original scientific ideas is concerned with his name. The question of life on other planets did not escape his attention either.

In the early 1820s Gauss became interested in geodesy, the mathematical study of Earth's actual shape. To increase the accuracy of surveying he invented the heliotrope, an ingenious instrument with a revolving mirror that by a simple sighting device can be moved by hand in such a way that the sunlight is always reflected in a certain direction.

He also came up with an ingenious way of using his heliotrope to contact the Moon's inhabitants, which reminds us of today's laser: 'With 100 banked mirrors, each sixteen square feet in area, one would be able to send a fine heliotrope light to the Moon,' he suggested. No one followed it up.

He had another idea. At that time it was believed that Martians could look at Siberia. To attract their attention, Gauss proposed clearing stretches of forest there to form a gigantic right-angled triangle with squares on each side; planting wheat in the triangle, and leaving squares of trees around it. He believed that his diagrammatic demonstration of Pythagoras' theorem would reveal our presence to our neighbours (how could Martian masterminds miss the mathematical message hidden in Pythagorean 'crop circles'?).

A few years later, Johann Joseph von Littrow, an Austrian physicist, worried about the poor visibility of Pythagoras' diagram in the dense Siberian forest, suggested that the right thing to do would be to dig giant ditches in the Sahara in the form of geometric figures such as triangles, squares and circles, fill them with water, and pour kerosene on top of the water and set them ablaze at night. Even this bright idea was not followed up.

The idea of a blazing Sahara as a beacon to attract Martians' attention appeared again in 1880 when French astronomer Camille Flammarion suggested that, if chains of light were placed on the Sahara on a sufficiently generous scale to illustrate Pythagoras' theorem, intelligent Martians might conclude that there was intelligent life on Earth.

These ideas certainly fired up the imagination of many 20th-century scientists. In 1920 Guglielmo Marconi said that he had received mysterious radio signals that he believed might have come from Mars or some other region of the universe where electrons are in vibration. He ignored these signals, saying: 'I'm concerned enough at present with business upon Earth.' The *International Herald Tribune* (14 February 1920) reported: 'Mr. Thomas A. Edison, commenting on the statement of Marconi that untraced wireless calls might come from Mars, stated that such a thing is possible.

"Existing machinery is able to send signals to Mars," said Mr. Edison. "The question is, have the beings there instruments delicate enough to hear us? They say Martians are as far ahead of humans as we are ahead of chimpanzees. If that is true they must have such apparatus."'

In 1924 David Todd, an American astronomer who like Edison believed in the superiority of Martian technology, convinced the US government to turn off high-powered radio transmitters on 23 August, when Mars came closest to Earth. On that day, when the two worlds were just 56 million kilometres apart, transmitters were turned off for five minutes before each hour, so that Todd could listen to Martian chatter. Martians were also aware of Earth's closest approach. On that day they decided to turn off their radio stations and sit in silence. Well, they already knew that 'turning off relaxing music and sitting in silence is good for your health' (L. Bernardi et al. in *Heart*, April 2006) – something that our researchers found decades later.

Todd did not give up his hope of communicating with Martians that easily. He suggested filling a 15-metre bowl with mercury at the bottom of an abandoned Chilean mine and placing a powerful light source at the focus of this natural parabolic reflector. This time he failed to convince authorities to act on his good advice.

Startling Martians with maths

1920 was a vintage year for scientific ideas about communicating with Mars. Pages and pages of *Scientific American* for that year are filled with suggestions not only for technical means of signalling but also about the 'language' of the message. 'Should we signal in English or in French or in what language?' asks a reader. 'Or perhaps the enthusiastic mathematicians who tell us, with delightful generality, that mathematics is the universal truth and that therefore the message will be such as to convey some fundamental mathematical idea, will deign to work out their scheme for us, and tell us with just what mathematical truth we are to startle the Martians?'

Someone suggested giving Martians an IQ test by sending two signals, then four, and if they answered back with eight signals we would know that Martians could multiply.

Not everyone was convinced of the existence of Martians. One reader wrote that if all the letters of the Lord's prayer were thrown into the air, there would be as much chance of the letters 'falling back into their proper places to print the prayer without an error as there is of there being inhabitants on the planet Mars with whom we might by any possibility communicate'.

These views did not deter the editors of *Scientific American*, a weekly magazine then, to publish a very long article, 'What shall we say to Mars?', on 'this rather unorthodox scientific matter' in the issue of 20 March 1920. The article described an elaborate scheme of writing a pictorial message (see Figure 6) in 'dots or dashes, of wireless or light, as the case may be', and then using these pictures to teach Martians a new language. Image d) shows, directly beneath the picture, a group of dots and dashes. They spell, in Morse code, the word 'man'. The article suggested that the following messages would be labelled 'hand' and 'head', and so on. This practice would compel the Martians to the conclusion that these marks were arbitrary symbols for the various objects pictured, and thus would help them in 'building up the interplanetary language'. In practice, the article said, 'we shall not use English words, but some language of more logical construction, like Esperanto'.

In 1942 British astronomer James Jeans suggested

attracting the attention of Martians by shining a group of searchlights towards Mars and emitting successive flashes to represent series of numbers. 'If, for instance, the numbers 3, 5, 7, 13, 17, 19, 23 ... (the sequence of prime numbers) were transmitted, the Martians might surely infer the existence of intelligent Tellurians [inhabitants of Earth].'

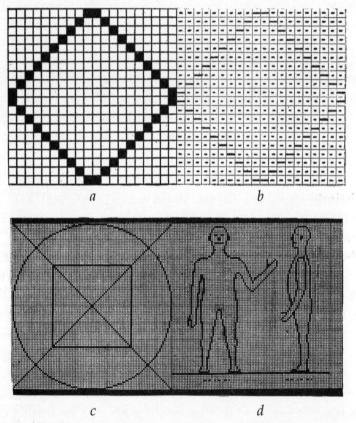

Figure 6. How dots and dashes can be used to send pictorial messages to Mars. a) Dots and dashes arranged to form a simple diamond; b) dashes of previous message plotted on quadrilled paper; c) more elaborate message, with the dots again eliminated; d) a still more complicated message, depicting a man. (Reprinted with permission from Scientific American, *20 March 1920)*

Say hello in prime numbers

Even today, SETI researchers do not expect to hear a hello from an ET from their speakers, but they hope to see their computers recording an intelligent pictorial or mathematical message. A series of short and longer transmissions representing dots and dashes could be used to send a pictorial message. A mathematical message could be sent as a prime number or as Fibonacci numbers: 1, 1, 2, 3, 5, 8, 13, 21, 34, 55, etc. (each successive term is the sum of the preceding two).

Fibonacci numbers have many interesting universal properties. For example, the ratio of successive terms (larger to smaller: 1/1, 2/1, 3/2, 5/3, 8/5 ...) approaches the number 1.618. This ratio is known as the golden ratio and is denoted by the Greek letter phi (Φ). Phi was known to ancient Greeks, and Greek architects used the ratio 1:Φ as part of their designs, the most famous of which is the Parthenon in Athens. Curiously, phi also appears in the natural world. Flowers often have a Fibonacci number of petals (look at the arrangement of florets on a cauliflower). The seeds on a sunflower are arranged in two sets of spirals. The ratio of the number of seeds in the two spirals is phi – and so is the ratio of your height and the distance of your belly button from the tip of your feet. Do the aliens have belly buttons?

Mathematicians are probably not interested in the answer, but they believe that mathematics is universal and that any intelligent aliens will understand the language of mathematics. 'Are the mathematics and the mathematical descriptions of physical phenomena used by an alien technological civilization on another planet likely to resemble what human beings come up with on Earth, even if the notation is different?' asks Murray Gell-Mann, the American theoretical physicist who introduced the concept of quarks.

He recalls an anecdote from his term as a visiting professor at the Collège de France in the 1960s. He was discussing Project Ozma with someone, and how communication with the aliens might take place: 'I suggested we might try beep, beep-beep, beep-beep-beep, etc. to indicate the numbers 1, 2, 3, and so forth, and then perhaps 1, 2, 3, ... 42, 44, ... 60, 62, ... 92, for the atomic numbers of the 90 chemical elements that are stable 1 to 92 except for 43 and 61.' The person replied: 'Wait, that is absurd. Those numbers up to 92 would mean nothing to such aliens ... Why, if they have 90 stable chemical elements as we do, then they must also have the Eiffel Tower and Brigitte Bardot.'

Gell-Mann replies by saying that the aliens will arrive at the same physical laws 'even if each of them has seven tentacles, thirteen sense organs, and a brain shaped like a pretzel'. He agrees that their mathematical notation is very unlikely to resemble ours. Most scientists also believe that the fundamental laws of science provide a common 'language' throughout the universe, since they hold true in any region of it.

Leopold Kronecker, a 19th-century German mathematician, had a different view: 'The integers were created by God; all else is the work of man.'

In the 21st century, Indian philosopher Sundar Sarukkai notes that even if numbers or counting may have a common genesis, mathematics on other worlds may differ considerably from ours. He provides two arguments:

First, there are no unique mathematical descriptions of physical phenomena. We extract physical concepts from these phenomena, which are then mathematically described. And the way we extract physical concepts from the world depends to a great extent on our physical and mental capacities. For example, if our eyes were sufficiently tuned to detect that space is curved, we could have known it much before Einstein! Thus, the physical description we extract from the world to which we match certain mathematical ideas depends to a great extent on our capacities of perception.

Second, the mathematics known to us arises from our interaction with our physical world. Thus, it depends not only on the nature of our physical world but also on the nature of our perception. We see the world in discrete terms, such as seeing objects as discrete entities. Moreover, the mathematics we create depends on the structure of our body and of our mind. There is no human mathematics without writing and there is no writing without fingers that do the writing. Would intelligent aliens who have no writing develop mathematics as we know it? I think the answer is no. For creatures which have different bodies and minds, there is no conceivable way that they would have the same mathematics as we do.

Although mathematics is not universal in the way scientists think of it, he believes, it does not preclude the possibility of using mathematics for communicating with aliens. 'This is primarily because of the capacity for mathematics to be taken into the folds of translation in very interesting ways,' he says.

A singing astrogram

Humanity's first deliberate message to other civilisations was sent from the Arecibo radio telescope in 1974. It was beamed towards Messier 13, a group of some 300,000 stars in the constellation Hercules 25,000 light years away. Carl

Sagan estimated that there is 'about a one in two chance of there being a civilization in Messier 13'. The message was transmitted in 169 seconds at a radio frequency of 2,380 megahertz (wavelength 12.6 centimetres). The strong signals containing the message, which were repeated many times, will be detectable by Arecibo-sized radio telescopes throughout the galaxy.

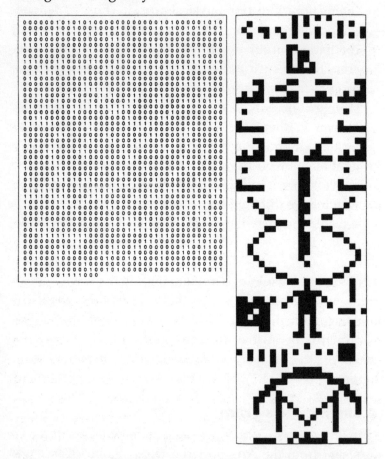

Figure 7. The 1974 Arecibo message to the stars in binary numbers (left) and in a bitmap image (right). (Images: National Astronomy and Ionosphere Center, Cornell University)

The message consists of 179 consecutive characters written in binary numbers, one of the simplest number systems with only two symbols, 0 and 1. In the actual transmission, 0 and 1 were represented by one of the two specific radio frequencies, and the characters were sent sequentially by shifting between these two nearby frequencies.

The message can be broken down into a grid, 23 characters by 73, which can be used to build a bitmap image (see Figure 7). The numbers 23 and 73 are prime numbers, and it is believed that these will help the aliens to arrive at the fact that this grid is the only correct way to interpret the message.

The opening line of the message, right to left, describes numbers 1 to 10 in binary form. This 'lesson' will help the receiver in deciphering the message. The next three lines indicate the atomic numbers of five chemical elements – hydrogen, carbon, nitrogen, oxygen and phosphorus – that are important in living cells.

Next, the message uses this information to describe the molecular components of DNA. The double spiral below shows the DNA molecules; and at the spiral's central core is the number 4 billion – the number of characters in the genetic code. The DNA is followed by the crude sketch of a human (it looks like the legendary Australian bushranger Ned Kelly in gumboots, according to a wit), establishing the connection between the DNA and the intelligent creature. To the right of the human is a line extending from the head to the feet, and accompanied by the number 14. This means that the height of the human is 176 centimetres, 14 times 12.6 centimetres – the unit of wavelength that is used to transmit the message. To the left of the human is the human population of Earth (4 billion).

The next line shows a sketch of the planets of the solar

system, with the Sun at the right and Earth displaced towards the human figure, suggesting that the third planet is the home of the creatures who sent the astrogram. The message is signed by the Arecibo telescope with its diameter of 305 metres shown as a multiple of 12.6 centimetres.

The message was developed by Frank Drake and his colleagues at Cornell University's National Astronomy and Ionosphere Center. The message was 'market tested' to see how easy it was to decipher. 'But a spot check in the *Nature* office revealed no intelligences (native or alien) that could decipher it,' reported the journal on 24 January 1975. 'Are interstellar IQ tests culture fair?' the journal wondered.

No codes, just the sounds of the seventies

Identical engraved plaques attached to Pioneers 10 and 11 spacecraft, launched in 1972 and 1973 respectively, were definitely culture fair. Each plaque measures 15 by 22 centimetres and is made of gold-anodised aluminium. These cosmic greeting cards were designed by Carl Sagan and Frank Drake and drawn by artist Linda Salzman Sagan.

On the plaque, a nude human male and female stand before an outline of the spacecraft coming from the third of nine planets around a star (see Figure 8). The star's location is shown with respect to fourteen pulsars and the centre of the galaxy as the radial pattern of dashed lines at left centre. The precise periods of the pulsars are shown in binary numbers to allow them to be identified. The heights of man and woman are shown with respect to the spacecraft, and the height of the female is also given as a binary number eight on the far right of the plaque.

Units of time and distances are expressed relative to the wavelength of 21 centimetres emitted by a hydrogen atom. This element is illustrated in the top left-hand corner in schematic form showing the hyperfine period of neutral hydrogen (a fraction of a nanosecond). If alien physicists can understand this information, they can unscramble the message. For example, they can work out that 8 times 21 centimetres is the height of the female earthling.

According to Sagan and Drake, the plaques provide sufficient information to our star in 250 billion stars, and one year, 1970, in 10 billion years: 'These plaques are destined to be the longest-lived works of mankind. They will survive virtually unchanged for hundreds of millions, perhaps billions, of years in space.'

Contact with Pioneer 11 was lost in 1995. The last signal from Pioneer 10 was received in 2003 and now it's on a course to visit the star Aldebaran in the constellation Taurus. This red star is 68 light years away and it will take Pioneer 10 roughly 2 million years to arrive. If there are any inhabited planets around this star, we can expect a thank-you card in 4 million years.

Voyager 1 and 2, launched in 1977, each carry a 30-centimetre gold-plated copper phonograph record containing 115 images and greetings spoken in 55 languages, selected to portray the diversity of life and culture on Earth. Each record is packed in an aluminium jacket together with a stylus and cartridge. It's designed to play images and sounds at $16\frac{2}{3}$ revolutions per minute. The record's image and sound are encoded in analogue, which has been superseded by digital coding. 'With hopeless gallantry, we built these messages to last a billion years with playback equipment that was obsolete before it got to Neptune,' laments E.C. Krupp of Griffith Observatory in Los Angeles. The Voyager spacecraft are now at the edge of the solar system and on track to a faraway radio station to play golden sounds of the seventies.

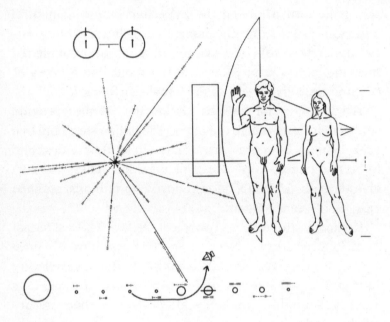

Figure 8. The gold-anodised plaque affixed to the Pioneer 10 and 11 spacecraft (launched in 1972 and 1973 respectively). (Image: NASA)

Message in light fantastic

Why, despite nearly five decades of listening, haven't we picked up any obviously intelligent radio signals?

Perhaps no civilisation is continuously broadcasting a powerful 'beacon' signal from a planet around any of the nearest 1,000 or so Sun-like stars. If such signals existed, they would have been picked up by powerful searches undertaken by SETI researchers in the past decade.

Or, we are searching such a small range of frequencies that it's like poking around the edges of the 'cosmic haystack' for the 'needle' of an alien signal. This haystack is estimated to contain 100 trillion radio channels, sky directions and other parameters. 'Earthlings are just getting into the game,' remarks Dan Werthimer, an American SETI astronomer. 'We can rule out hardly anything.'

Or, ETs are quite imaginative and they may have moved to technologies far superior to radio signals. Such as lasers.

American physicist Charles Townes, who shared the 1964 Nobel Prize for the invention of lasers, realised around the time that Drake aimed his telescope at distant stars that ET civilisations could just as easily exploit the optical and infrared portion of the spectrum as the radio portion. Decades passed before laser technology had advanced to the point where powerful lasers capable of sending inter-stellar messages could be made.

Lasers produce intense light of usually very pure frequency or wavelength, which is emitted in an extremely tight beam. This beam can be easily focused on a target as small as one millionth of a millimetre. High frequencies produced by lasers, several hundred million megahertz, give them enough band-width for a very high rate of trans-mission of information. Thus the laser has a major advan-tage over the millions of radio channels available for broad-casting: it can transmit a whole movie, if the aliens are into movies, in seconds – much better than simply asking 'How are you, earthlings?' by a radio wave. They offer another advantage. If we do receive a laser signal, it will be much easier to locate the source, as lasers are uni-directional.

SETI enthusiasts began searching for laser signals in the early 1990s, but they have not yet beamed any laser messages to other planets. Paul Horowitz, a Harvard University physicist, is a veteran of laser searches. He notes that lasers have now grown so powerful that a laser beam could be 5,000 times brighter than the Sun. Similarly, optical telescopes fitted with photo-detectors have now become so powerful that they can record extraterrestrial pulses of laser light only a few billionths of a second (nanoseconds) in duration. Using a 1.5-metre telescope at Oak Ridge Observatory, Horowitz and his colleagues have made more than 16,000 searches for nanosecond pulses from 13,000 Sun-like stars. No ET seems to be beaming a copy of *Encyclopaedia Galactica* in our direction.

Townes now believes that flashes of light from planets around stars within 50 light years could even grow bright enough for the naked eye to see. When it does happen, all you have to do to search for ETs is to relax in a lounge chair and look up at the sky.

Message in a bottle

If aliens want to contact us, they are not likely to send their *Encyclopaedia Galactica* 'we are here' message by a pulse of radio waves or lasers, but in an old-fashioned way – by a written note (in a perfumed envelope, we hope). After analysing the energy efficiency of long-distance communication, American researchers Christopher Rose and Gregory Wright suggest that 'our initial contact with extraterrestrial civilizations may be more likely to occur through physical artefacts – essentially a message in a bottle – than via electro-magnetic communication'. They explain their calculations with the concrete example of the Voyager spacecraft, which is carrying a recording containing about 1 billion bits of information. If it were carrying three DVDs with 100 billion bits of information, they calculate, it would be a more energy-efficient way of sending this amount of data to someone 200 light years away than an Arecibo-to-Arecibo radio communication.

Unless the message is short or a prompt reply is desired, sending a message inscribed on some material requires less energy per bit of transmitted information than sending it by radio waves, the researchers say. 'If energy is what you care about, it's tremendously more efficient to toss a rock.' Beams of radiation spread out as they travel through space, diluting the signal and wasting energy. An inscribed physical object, on the other hand, is not 'diluted' as it travels through space. Radio announcements last momentarily unless repeated continuously. Physical objects last forever, almost.

A message in a bottle could be waiting for us in our own planetary backyard, like the obelisk left on the Moon by aliens in Arthur C. Clarke's *2001: A Space Odyssey*. Any artefacts from aliens are likely to be orbiting the Sun or a planet, or resting somewhere on a planet, moon or asteroid. Rose and Wright advise that 'carefully searching our own planetary backyard may be as likely to reveal evidence of extraterrestrial civilizations as studying stars through [a] telescope'.

The *New York Times* is among those who are sceptical of receiving a bottled message from aliens: 'The only hitch is that such a message, travelling at a crawl by space standards, might have been sent 30 million years ago, and who knows whether the senders would have moved on or died off in the eons since. Best not to give up on electromagnetic messages entirely.' ET, please phone us.

Sorry, the number's not yet connected

For interstellar communication, messages should travel as fast as possible, preferably at the speed of light. That's why we expect extraterrestrials to communicate by radio or other electromagnetic waves. But some interstellar civilisation may be using neutrinos instead of electromagnetic waves for communication. Neutrinos are elementary particles with no charge, and are nearly massless. They travel at or near the speed of light. They are everywhere – trillions of them pass through our bodies each second – but they cannot be seen and rarely interact with matter. They are the most elusive of all the elementary particles we know, yet scientists have found ways to detect these ghost-like particles.

A massive neutrino telescope is being built at the South Pole to trap neutrinos from deep space. These originate in black holes or crashing galaxies which give them enough energy to travel billions of light years. The telescope, dubbed IceCube, will sample neutrinos from the sky in the northern hemisphere. Earth will act as a filter to exclude neutrinos originating from the Sun. When completed by the end of the decade, the telescope will house 60 basket-ball-sized optical detectors in each of 70 2.4-kilometre-deep holes drilled into the Antarctic ice. These 4,200 optical detectors will occupy one cubic kilometre of ice, hence the name IceCube. The detectors will record tell-tale signatures of neutrinos when they pass through Earth from the northern hemisphere and collide with other atoms. This rare collision produces another elementary particle called a muon. The muon leaves a trail of blue light in its wake which allows scientists to trace its direction back to the point of origin of the neutrino. IceCube may even one day eavesdrop on alien teens talking on their nifty neutrino mobiles.

Walter Simmons and his colleagues at the University of Hawaii in 1994 were the first to suggest that extraterrestrial civilisations might be using neutrinos to broadcast information throughout the galaxy. They came to this conclusion after reasoning that any advanced civilisation is bound to have exacting time standards. A civilisation that has spread through our galaxy might send 'timing pulses' over interstellar distances in order to synchronise clocks on different worlds. The decay of a particle known as a Z boson into a neutrino and an anti-neutrino takes just one thousand billion billionth of a second, making it the fastest process known in physics. Neutrinos have other advantages over photons (which make up electromagnetic waves): they are not blocked by interstellar dust or dispersed by ionised gas, and are brighter than photons. A very bright source of neutrinos could transmit precise timing pulses across many thousands of light years. These neutrino signals will be quite distinctive, as they will come from a specific direction in the sky and have a very well-defined energy.

No 18th-century scientist could have dreamt of sending sounds and pictures around the planet by electromagnetic waves. What about a civilisation trying to communicate with us via a medium as yet undreamt of? Our neutrino communication lines are now open, but they might be sending 'smoke signals' we are not yet ready to decipher.

Look for 'smoke signals' in space

These 'smoke signals' might be in the form of giant solar sails in space, if we believe Luc Arnold of the Observatoire de Haute-Provence in France. 'Artificial structures may be the best way for an advanced extraterrestrial civilisation to signal its presence to an emerging technology like ours,' he says. These planet-sized structures could be like light-weight solar sails or other very low-density structures specially built for the purpose of interstellar communication.

When a planet crosses in front of its star as viewed by an observer, the event is called a transit. Transits produce a small and periodic change in the parent star's brightness. All transits of the same planet produce the same change in brightness and last the same amount of time, thus providing a highly accurate method of detection. If planet-sized artificial objects exist around other stars, they would always transit in front of their star for a given observer in space. These objects can be spotted by a new generation of space-based telescopes such as NASA's Kepler Mission. Kepler is specifically designed to detect Earth-sized or smaller planets in our galaxy by studying their transit.

After studying several simulated artificial transits, Arnold has determined the characteristic transit signals of differently shaped objects. A Jupiter-sized triangle, for example, will produce specific transit signals which are different from the transit signals of a Jupiter-sized planet. Multiple objects produce remarkable transit signals because of their 'on again – off again' nature of light. Such an observation would clearly show that the transiting objects are artificial. The best way to visualise this is to imagine a flashlight moving behind a lowered window blind.

Arnold says that a civilisation would rather build a series of small objects to generate multiple transits rather than a large single object for saying 'Hi, we're here' to the whole galaxy. He imagines, for example, that if eleven objects were orbiting a star in groups of 1, 2, 3 and 5 – the first five prime numbers – the mathematical pattern of the dimming of the star's light would clearly represent a message from the aliens.

'Transit of artificial objects also could be a means for interstellar communication *from* Earth in future,' he says. His advice to future generations: to have in mind, at the proper time, the potential of Earth-sized artificial structures in orbit around the Sun to say hello to intelligent extra-terrestrials. They may not have the technology to decipher messages sent by radio waves, lasers or neutrinos, but they can't miss 'smoke signals' in the sky.

The message is in our genes

Forget radio waves, lasers, neutrinos or giant structures around stars – the most likely place to find an alien message is in our genes. Paul Davies believes that the aliens have left their message in the junk DNA in our genes. The term 'junk DNA' is used for those portions of DNA which appear to have no specific function. Scientists have recently discovered large sequences of junk DNA which have not changed over the last 75 million years of evolution.

A DNA sequence can contain more than a million base pairs, which is enough for a decent-sized book. The DNA need not be the last word from ET, says Davies. 'Rather, it could tell us how to download the entire contents of *Encyclopaedia Galactica* by conventional radio or optical techniques.'

Why can't the aliens broadcast the encyclopaedia now? Davies says that it's inconceivable that ET would beam signals at our planet continuously for untold aeons merely in the hope that one day intelligent beings might evolve and decide to turn a radio telescope in their direction. Nor does he entertain the idea of the aliens leaving an obelisk at our doorstep, because such an object would be subject to the vagaries of tectonic activity, glaciations and other natural turmoil.

From his point of view, 'a better solution would be a legion of small, cheap, self-replicating machines that can keep editing and copying the information and perpetuate themselves over immense durations in the face of unseen environmental hazards'. Nothing beats the living cell. He speculates that ET might have inserted a message into the genomes of terrestrial organisms, perhaps by delivering carefully crafted viruses in tiny space probes to infect host cells with message-laden DNA.

To read this cosmic greeting card, Davies suggests

displaying a sequence of junk DNA bases as an array of pixels on a computer screen. Patterns that stand out, such as prime numbers, would be a clincher.

ET, please call Earth in Lincos

Lincos – short for the Latin *lingua cosmica* – is a universal language developed in 1960 by Dutch mathematician Hans Freudenthal. It's not a spoken language, but a complex mathematical scheme for use in interstellar radio communication with extraterrestrials.

Freudenthal suggested that first we should 'communicate facts which may be supposed to be known to the receiver'. Obviously, these facts must be concepts that are universal. Humans learned to count before they learned to write; mathematical concepts are believed to be universal. The first lesson in Lincos starts with a simple pattern of radio pulses to establish symbols for natural numbers in binary notation and basic arithmetical processes: addition, subtraction, multiplication and division. Each symbol is explained by symbols that came before it, so you don't have to know anything except pure mathematical concepts to understand it. Freudenthal believed that 'decoding Lincos would be an easy job'. However, ETs must have the technology to receive radio signals and measure their frequency. Lincos phonemes that make up words are transmitted as radio signals of varying duration and frequency.

The first interstellar message in Lincos was broadcast in 1999 from a 70-metre radio telescope in Evpatoria, a small town in Ukraine. It was aimed at stars 51 to 71 light years away in a region of sky called the Summer Triangle. The signal was 100,000 times stronger than a TV broadcast. The 400,000-bit message was much longer than the 1,679-bit Arecibo message broadcast in 1974.

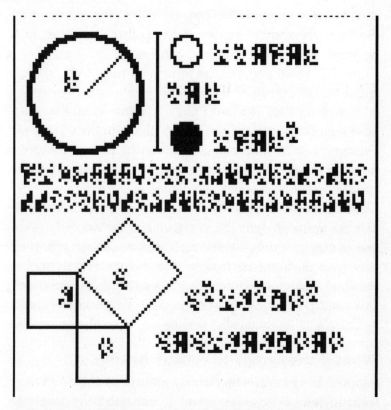

Figure 9. A geometry lesson for aliens. Euclid might not have understood it, but the aliens will. The top section introduces radius and area. In the middle, pi is displayed with 1,241 billion digits. The bottom section shows the graphical representation of Pythagoras' theorem. (Image: Yvan Dutil and Stéphane Dumas)

The message was developed by physicists Yvan Dutil and Stéphane Dumas, who work for the Canadian government. SETI is a hobby for both of them, and the message was developed as a private project. They believe in active SETI – the active search for extraterrestrials. 'In classic SETI you listen as hard as you can, in active SETI you shout as loud as you can,' says Dutil. 'By the way, to be honest, active

SETI is the true classic SETI, since it was first proposed by the German scientist Gauss at the end of the nineteenth century.'

In 2003 Dutil and Dumas improved their 1999 'shout', and it was broadcast on the same day from radio telescopes in Evpatoria and Roswell, New Mexico. Unlike a pure Lincos message, their message also uses graphics. The 1999 message was an interstellar fax of 23 pages; the 2003 message is one very long page. In fact, it's a lesson in mathematics, physics, biology and cosmology (see Figure 9). It will be difficult for visually challenged aliens to decipher it, as some form of sight is a prerequisite. We can only hope that if they can develop the technology to receive it, they may have the ability to 'see' it. Will they be able to understand it? Easily, if they have an Alan Turing among them. Dutil and Dumas say that the message will decipher itself, as it also teaches those who receive it how to crack the code.

What if they want to remain hidden?

Suppose ETs have received the Evpatoria message and they want to send a thank-you card to earthlings for teaching them Pythagoras' theorem and other such earthly things. But they don't want their message traced back to them; they want to remain hidden.

Walter Simmons and his University of Hawaii colleague Sandip Pakvasa believe that they know how ETs can do it. They say that in cryptography a random signal is added to a message to make it indecipherable by anyone other than the sender and the intended recipient. Both the sender and the recipient know the random signal. When the signal is subtracted from the message, it becomes decipherable. ETs can achieve the same effect by splitting their message into two parts and sending them in opposite directions to two mirrors located far from the sender's planet. The mirrors reflect the signals to the intended receiver. The message is revealed when the two parts are combined.

The researchers then solve the problem of how the two parts cannot be decrypted individually, either before or after they are combined and read. They suggest that if the message is tiny enough to fit on a microscopic particle, Heisenberg's uncertainty principle would make it impossible to read the message and the direction simultaneously. The principle says that it's impossible to determine exactly the position and momentum of a particle simultaneously. By reading the message, all the information about its origin will be lost.

There is another problem: how two signals can carry an image without leaving any trace of its presence in either signal. The researchers expect a quantum effect known as 'entangled particles' to help ETs. This effect allows a pair of photons to be connected at a distance and to influence each other instantaneously, whether they are in the same room or at opposite ends of the galaxy. The events affecting one photon will also modify the other photon. Einstein never liked this 'telepathic connection' between particles. He called it 'spooky action at a distance'. The Hawaii duo expects quantum entangled photons to unite to form the microscopic image of the alien message. Scientists have recently shown that entangled particles can indeed be used to carry an image.

ETs may be using this strategy to send interstellar messages to other Galactic Club members, if for some unknown reason they do not want to reveal their presence to us. We do not have the technology to look for such alien signals today, Simmons and Pakvasa say, but we may have it in a decade or so.

Terrestrial principles may be alien to the aliens

Richard Dawkins, world-renowned evolutionary biologist, claims that 'the Darwinian law may be as universal as the great laws of physics', and complex structures found anywhere in the universe are of living origin. Elling Ulvestad, a medical researcher at Haukeland University Hospital in Norway, has applied this 'theoretical framework' to arrive at an intriguing hypothesis: extraterrestrial radio signals are not necessarily generated by the aliens.

His argument runs something like this: We can envision extraterrestrial radio signals as artefacts generated by intelligent life on other planets (hypothesis H_1). Although the theory of evolution is opposed to the design argument for explaining biological diversity, it's not opposed to using it for explaining artefacts. As such, the design argument is valid for scientific inference. Radio signals could therefore be looked upon as valuable signs of meaning in the universe. However, when regarded as a scientific hypothesis, we have little reason to believe H_1 rather than the opposite hypothesis – that the signals are not generated by an alien civilisation (H_2). Since there is no valid data to support either hypothesis, both hypotheses have an equal chance of being right: probability (signal/H_1) = probability (signal/H_2).

Why do we then believe that the likelihood of H_1 is greater than H_2 – that signals are coming from intelligent life? It's our Earth-centric background knowledge that inclines us to believe it, explains Ulvestad.

He says that our ever-increasing technological sophistication and an almost obsessive wish to learn whether we are alone in the universe may well turn astrobiology into the major scientific enterprise of this millennium. The laws of physics may be universal, but astrobiologists have no valid pre-understanding of their subject matter. 'Since there is no assurance that life beyond Earth will be Earthlike, terrestrial biological principles and data do not necessarily provide valid information when inferring life in other worlds. Paradoxically, extrapolation of terrestrial principles to other worlds may preclude any chance of finding new principles of life.'

Ulvestad solves this paradox by suggesting that unprejudiced information about alien life forms can be gained only by applying biosemiotics (*bio*, life; *semion*, sign), the study of signs, of communication, and of information in organisms. He says that semiosis, any form of activity or process that involves signs, is 'a universal characteristic of living organisms, because without semiosis there can be no recognition ... Semiotic communication involves the sign, the object that the sign refers to, and the interpretant. For something to be a sign it must be understood as such – a sign is a sign only in context. Signs must be interpreted in relation to each other in a context, otherwise they may not even be acknowledged as signs.'

SETI researchers use radio signals, he continues his argument, because of the practical convenience of radio waves. Their Earth-centric presumption expects that aliens are equipped with receivers among their sense organs that respond to the same auditory signals that humans do. Furthermore, since the reception of any message is dependent on prior knowledge of the possibilities, he says, it is expected that the aliens have an evolutionary history similar to ours. 'I find both assumptions incomprehensible, and consequently find the utilization of radio waves as a means for contact with extraterrestrial intelligence dubious also from an evolutionary angle.'

Hello Earth!

Imagine this: Scientists have received an alien radio signal. After a painstaking process of checking and verification, they believe that the signal is really from an extraterrestrial intelligence. What happens now? A five-minute hypothetical on this Earth-shattering discovery.

Importance

'It would certainly be the greatest discovery of all time, eclipsing the findings of Newton, Darwin and Einstein combined,' comments Paul Davies. 'The knowledge that we are not alone would affect people's psyche and totally transform our world view.'

'The discovery would remove the final cloak of superiority,' adds Seth Shostak: 'We would know that we are neither culturally nor intellectually supreme, but simply one society among many. We would no longer be special.'

Secrecy

Will there be an *X-Files*-type cover-up or pressure from the authorities to classify information? No, emphasises Shostak: 'The real problem is not that an alien transmission would, or could, be kept secret. Rather, scientists worry that a promising signal would generate so much excitement that it would be announced prematurely, only to prove later to be of earthly origin.'

Interpretation

Once the signal has been proved to be from an intelligent civilisation, scientists' top priority would be to interpret the message. If they are not successful in interpreting it without any ambiguity, American author John Casti (*Paradigms Regained*, 2000) is worried that it may 'maximise the opportunity for malicious exploitation, as the public would be distracted from the scientific attempts to figure out what the message really means by the sideshow of events centring on those who claim to know the meaning or claim that the SETI community and/or government knows the meaning and will not divulge it'. This scenario, he says, presents enormous opportunities for rumour, exploitation and fear.

Impact

The impact of the message on our society would depend upon its correct interpretation and its intention: whether the message is a simple greeting, 'Hello Earth!'; instructions to download the latest *Proxima Centauri Idol* show and such other goodies by radio waves; a command to pack up all our uranium in large crates and leave them in the middle of the Sahara desert so that they can be picked up by an alien spaceship; or a declaration of invasion in the style of *The War of the Worlds*. Most likely it would be a passive greeting directed at our solar system, not specifically at Earth, suggesting that, like us, the sending civilisation is scanning the galaxy for intelligent life.

Sociological studies suggest that the message would lead to confusion and excitement, with a desire by individuals to 'know more', but little panic or hysteria, says a publication from the SETI Institute. For some sceptics, the news could be disturbing.

The proof of the existence of intelligent life elsewhere in the universe might reinforce the religious beliefs of some people, but it might create problems for those who take their religious texts literally. Some religious groups might reject the idea altogether, but major religions are expected to cope well with this discovery, as they have with other scientific discoveries.

Reply

There is no rush to reply. The message might have taken hundreds of light years – 100 light years is the most optimistic estimate by SETI researchers – to travel to Earth. Our reply would take centuries, even millennia, to reach them. The sending civilisation might even become extinct (and their planet turned into a *Planet of the Apes*) by the time the reply reaches them.

As there is no international treaty defining the process of consultation (see page 192), the discovery might lead to political bickering among nations.

Should we reply?

George Wald, the famous biologist who won the 1967 Nobel Prize for physiology or medicine, and who believed that life probably originated on other worlds, was worried about humankind's self-esteem when he remarked in 1974: 'I can conceive of no nightmare so terrifying as establishing such communication with a so-called superior (or if you wish, advanced) technology in outer space.'

Zdeněk Kopal, professor of astronomy at Manchester University from 1951 to 1981, put it bluntly: 'If the cosmic telephone rings, for God's sake let us not answer, but rather make ourselves as inconspicuous as possible to avoid attracting attention!'

In 1976 Martin Ryle, Britain's Astronomer Royal from 1972 to 1982 and sharer of the 1974 Nobel Prize for physics, attracted worldwide media attention when he suggested that we might become the aliens' lunch. In a letter to the International Astronomical Union, he urged astronomers not to communicate with extraterrestrial civilisations. It was cosmic folly to reveal our existence and location to them, he warned. For all we know, he said, 'any creatures out there are malevolent or hungry', and once they knew of us 'they might come to attack or eat us'.

Frank Drake replied to Ryle's warning in the *New York Times* (22 November 1976): 'Put another way, Sir Martin Ryle has raised the question: Should mankind hide in this obscure corner of the tiny solar system attached to a minor star? ... The universe seems too rich to require an advanced race to look hungrily on Earth's meager patrimony.'

Our planet is a long, long way for the little green men. Even if they know that we're here, interstellar distances make it difficult for us ever to meet an alien. However, replying to a phone call is good manners. Hello ET!

What should the reply say?

Douglas Vakoch, a resident psychologist at the SETI Institute, is not waiting for a message. He's already working on a reply. His job is to design messages that are intelligible to extraterrestrials. If we ever get a message from them, what should be the focus of our reply? Altruism, replies Vakoch. Evolutionary biologists consider altruism – the self-less concern for the well-being of others – central to group survival; and they believe it may be widespread among extraterrestrials, as they would also have developed it as a survival technique. 'If they're right,' says Vakoch, 'altruism would be a good starting point for interstellar messages describing central concepts of terrestrial biology, behavior, and morality.'

It's not easy to convey the complex notions of altruism without any ambiguities to someone who doesn't speak our language. But Vakoch is confident that reciprocal altruism can be described through humans interacting. 'As an example,' he says, 'three-dimensional animation sequences could serve as "modern-day morality plays".'

'Perhaps the best clues that we are attempting to communicate altruism would come from the act of transmitting itself,' he adds. 'With no direct payoffs for senders, transmitting a celestial "message in a bottle" would convey something about our wishes to benefit extraterrestrials as well as future generations of humans, who might some day receive a reply of their own.'

Vakoch's message won't be limited to altruism. He is also working on a way to communicate the broad moral, ethical and religious principles to extraterrestrials. 'When people were deciding what to include on the Voyager spacecraft, they deliberately decided to exclude religious material,' he says. He wants to make sure that our next message tells the aliens how and what humans worship.

CHAPTER 8

Questions

What happens after we discover microbial life on another planet?

Most scientists believe that if life is discovered on a solar system planet or moon, it will be microscopic – perhaps on the surface of Mars or in a sea beneath the icy crust of Jupiter's moon Europa. European space scientists Paul Clancy, André Brack and Gerda Horneck say that such a discovery could fit three scenarios.

First, the life form is very similar to terrestrial life. Second, the life form is different from terrestrial life but is still based on a DNA/RNA/protein scheme. Third, a life form that is radically different from terrestrial life. This scenario presents the greatest challenge to scientists. This life form could still be carbon-based but might depend on geothermal energy emitted from a planet's interior instead of solar energy via photosynthesis. Or it could be silicon-based, or even with a completely different and unimaginable chemistry.

The trio says that the discovered life object, which may be a fossil or a living micro-organism, will not be the endpoint of the search: 'Rather it will be the starting point of a whole new era of scientific research.' They say that scientists would make their studies in a number of contexts. Scientists would like to know the *geographical location* – for example, if it's on Mars, whether in or out of a crater, in a mountainous or plain region – and the *geological context* – such as the type of rock – to plan new searches. The next stage of their research, the *biological context*, will focus on the study of the biology of the object, such as the structure of its cells, its metabolism and similarities with terrestrial organisms. The *geochemical context* will relate to the chemistry and mineralogy of the sample in which the object is found.

Studies in the *water context* will try to find the answer to the important question: What is the extent to which liquid water, either past or present, has been associated with the life-markers that have been discovered? The answer will indicate the importance of water for the genesis and survival of life. 'Has the case been overstated or not?' they ask. They consider the *contamination context* the most important aspect of the scientific study of the discovered life object; and emphasise that strenuous efforts will need to be undertaken to absolutely ensure that the object is not contaminated by humans or any robotic missions.

If the life object is a living micro-organism, Clancy, Brack and Horneck say that 'this would have impact not only on the daily news-chatter and preoccupations of Earth-bound humanity but also on the halls of serious science and philosophical communities, institutions and debate'. They claim that the scientific and technology communities are probably evenly split, but still a good 50 per cent tend to be hostile to the idea of life beyond Earth: 'So the discovery would settle all that and cause a real sensation.'

Could intelligent aliens be as small as nanobots?

Nano-robots or nanobots are hypothetical intelligent machines too small to be seen by the naked eye. They were envisioned by Eric Drexler in his book, *The Engines of Creation* (1986). Drexler, who is sometimes called the father of nanotechnology, imagined these tiny robots as self-replicating. Like biological cells, they would be able to make copies of themselves. In theory, they could build anything as long as they had a ready supply of the right kinds of atoms, a set of instructions and a source of energy.

Richard Smalley, who won the 1996 Nobel Prize in chemistry, dismisses the idea, saying that 'self-replicating, mechanical nanobots are simply not possible in our world'. To achieve their aim, nanotechnologists must provide these nanobots with 'magic fingers'. Within the constraints of a space of one nanometre – the size of nanobot – manipulation of atoms is not easy, because the fingers of a manipulator arm must themselves be made out of atoms. 'There just isn't enough room in the nanometre-size reaction region to accommodate all the fingers necessary to have complete control of the chemistry,' he says.

Let's apply this 'futurist's dream', as Smalley calls it, to intelligent aliens – aliens capable of building a technological civilisation. Could they be so small that they are almost invisible? By Smalley's logic the answer is a definite no. They won't be able to build anything. However, on a large planet where the pull of gravity is strong, it is possible that any life form may not be very tall. Therefore, aliens of the size of a large insect are possible. They could possibly have highly intelligent brains. With their 'magic fingers' they could build small robots that could build still larger robots and so on, and thus could have a technological civilisation far more advanced than our own.

What the law says about alien life forms

At present, extraterrestrial life forms, primitive or intelligent, have no rights under any national or international law.

The 1989 'Declaration of Principles Concerning Activities Following the Detection of Extraterrestrial Intelligence', approved by many international bodies such as the International Academy of Astronautics, the International Astronautical Federation, the International Institute of Space Law and the International Astronomical Association, states nine principles. In brief, they are:

- A signal or other evidence should be verified by its detector. If it can be confirmed, it should be treated like any other unknown phenomenon.

- All organisations parties to the Declaration should be notified so that they can run their own checks.

- If the discovery appears to be credible, national authorities and the rest of the scientific communities should be informed.

- Information of a confirmed detection should be disseminated promptly, openly and widely.

- Scientific data should be made available to the scientific community.

- The discovery should be properly recorded and monitored.

- If the detection is in the form of an electromagnetic signal, the appropriate frequencies should be protected by the International Telecommunication Union.

- No reply should be made until international consultation has taken place.

- If credible evidence of extraterrestrial intelligence is established, a committee of experts should be formed to analyse all data collected in the aftermath of the discovery, and also to provide advice on the release of information to the public.

According to Francis Lyall, a law professor at the University of Aberdeen, the Declaration is not binding in law. If signatories decide not to abide by it, 'at present there is no remedy by way of law, or any enforcement mechanism other than the disciplinary mechanism of scientific institutions.' His legal opinion is that it's not too early to consider formulating a United Nations treaty on the Declaration. 'Ideally, another entry in the catalogue of the UN Space Treaties would be the best,' he advises.

The United Nations-sponsored Outer Space Treaty states that 'the exploration and use of outer space ... shall be the province of all mankind'. Article 9 of the Treaty states that space explorations shall avoid 'harmful contamination' of Earth and other planets. All NASA missions are required to follow planetary protection procedures to avoid transport of Earth microbes to a planet, moon or asteroid. The procedures include careful cleaning and sometimes heat sterilisation of spacecraft. Similarly, precautions are taken against alien microbes hitch-hiking on space missions returning to Earth.

In 1969 the United States federal authorities passed a law that made contact between a US citizen and an extraterrestrial or their vehicles strictly illegal. The penalty for this 'extraterrestrial exposure' was maximum imprisonment of one year or a maximum fine of $5,000, or both. Curiously, the law was enacted on 16 July 1969, the day Apollo 11, the first manned mission to the Moon, was launched. Were the Feds worried about aliens hitch-hiking to Earth with Neil Armstrong, Michael Collins and Edwin 'Buzz' Aldrin? The law was repealed in 1991.

Whether they hitch-hike on earthlings' space missions or come by their own flying saucers, extraterrestrials have no protection under any local law. One can only hope that they have enough intelligence to buy interplanetary travel insurance.

How would inquisitive aliens find us?

Extraterrestrials would probably find us by applying the techniques employed by our own planet-hunters (see page 99). Once they found Earth, they would look for astronomical bio-signatures (signs of life visible from space) to determine if there was any life on their newly discovered planet.

If you used a Bunsen burner in your school days, you will find it easier to understand the science behind astronomical bio-signatures. The Bunsen burner was developed in 1855 by Robert Bunsen, a great German teacher and experimenter. Bunsen was a friend of Gustav Kirchhoff, who was a professor of physics at Heidelberg. Bunsen and Kirchhoff together developed the first spectroscope, a device used to produce and observe a spectrum. In 1860, Kirchhoff made the momentous discovery that, when heated to incandescence, each element produced its own characteristic lines in the spectrum. This means that each element emits light of a certain wavelength. In a leap of intuition he went even further: anything any atom emits it must also absorb. Sodium's spectrum has two yellow lines (wavelengths about 588 and 589 nanometres). The Sun's spectrum contains a number of dark lines, some of which correspond to these wavelengths. This means that sodium is present in the Sun. Scientists now had a tool they could use to determine the presence of elements in stars. According to Isaac Asimov, Kirchhoff's banker was not impressed with his ability to find elements in the Sun. 'Of what use is gold in the Sun if I cannot bring it down to Earth?' he asked. When Kirchhoff was awarded a gold medal for his work, he handed it to his banker and said: 'Here's gold from the Sun.'

If inquisitive aliens have scientists as smart as Bunsen and Kirchhoff, they will know how to mine gold from Earth by passing light through a spectroscope. From the resulting spectrum they will find out that the Earth's atmosphere has oxygen, water, carbon dioxide and ozone, suggesting the presence of life. If they find out that the atmosphere is oxygen-rich, they will know that photosynthetic life constantly replenishes the supply.

If they detect chlorofluorocarbons (from air-conditioning and refrigeration systems), methane (from sewage, rice cultivation, leaks from natural gas distribution and burps of domestic livestock) and smog (from fossil fuels) in the atmosphere, they will also know that their newly discovered planet is home to a high-tech society (which is intelligent enough to destroy itself by global warming).

If they have large radio telescopes, something like the Arecibo, they will receive radio waves from Earth, but they might find it difficult to distinguish these waves from those emitted by larger planets and the Sun. The signals from early television shows of the 1930s have so far travelled around 80 light years, too short a distance to reach another civilisation. Besides, these signals can be detected only if alien eavesdroppers tune to the right frequencies.

What would Earth look like to them?

Would they see the Great Wall of China? In 1938, well before the advent of space flights, American adventurer and intrepid traveller Richard Halliburton claimed in his *Second Book of Marvels: The Orient* that the Great Wall is the only man-made object visible to the unaided human eye from the Moon. This started the popular urban myth. The only thing you can see from the Moon, says NASA astronaut Alan Bean, is a beautiful sphere, mostly white (clouds), some blue (ocean), patches of yellow (deserts), and every once in a while some green vegetation. From the window of a spacecraft in a low orbit of Earth you can see the Great Wall, the Great Pyramid of Giza and many other man-made objects, if you know where to look and the weather is right. No artificial object is visible from the window of the International Space Station which orbits at about 360 kilometres up; however, you may spot big cities such as London, New York and Beijing in the daytime if you were good in your geography class.

The first-ever picture of Earth from Mars shows Earth blue and beautiful against the deep darkness of space. The picture was taken in 2003 by the Mars Global Surveyor spacecraft from a distance of 139 million kilometres. The most famous picture of Earth was taken in 1990 by the Voyager spacecraft from the edge of the solar system at a distance of 6.4 billion kilometres. In his book *Pale Blue Dot: A Vision of the Human Future in Space* (1994), Carl Sagan used the picture as a metaphor for the insignificance of our world in comparison to the cosmos. From a distance of 100 or more light years, aliens would see the 'pale blue dot' in their sky only if they had unimaginably powerful optical telescopes.

Are alien skies blue?

Why is the sky dark at night? This deceptively simple question has puzzled astronomers for centuries. And no, the
answer is not 'because at night the Sun is on the other side
of the Earth'. In 1823 Heinrich Olbers, a German astronomer, pointed out that if there were an infinite number of
stars evenly distributed in space, the night sky should be
uniformly bright – with a surface brightness like the Sun's.
He believed that the darkness of the night sky was due to
the absorption of light by interstellar space. But he was
wrong. Olbers' question remained a paradox until 1929,
when astronomers discovered that the galaxies are moving
away from us and the universe is expanding. The distant
galaxies are moving away from us at a speed so high that it
diminishes the intensity of light we receive from them. In
addition, this light shifts slightly towards the red end of the
spectrum. Red light has less energy than blue light. These
two effects significantly reduce the light we receive from
distant galaxies, leaving only the nearby stars which we see
as points of light in a darkened sky.

Why is the sky blue? This can be explained by the
scattering of light by molecules in the atmosphere. This
effect, known as Rayleigh's scattering, scatters shorter
wavelengths best. Therefore, blue light (wavelength about
475 nanometres) scatters more efficiently than red light
(about 650 nanometres). There is another reason. Our eyes
are sensitive to electromagnetic radiation in a very short
range. This 'visible light' range is from 400 to 700 nanometres (violet through red). Even within this range, we
don't perceive all colours equally. If our eyes were equally
sensitive to all colours, we would see the sky as violet (400
nanometres), as this colour is scattered more than the blue.
But our eyes are more sensitive to blue and green than

violet, so to us the sky is blue. We can only speculate how alien life forms, or even other life forms on Earth, would perceive the terrestrial sky.

The Moon and Mercury have no atmosphere; their skies are always black. On Mars, Rayleigh's scattering is very small, and during the day Mars' sky is yellow-brown. Around sunrise and sunset the sky is pinkish-red. The sky on Venus is yellow-orange. For other planets we don't have any space-probe data from within their atmospheres: Saturn and Neptune's skies are probably blue, and Uranus' sky is probably greenish-blue.

On extra-solar planets, the night sky would look very different to someone from Earth. The Sun would be visible to the naked human eye up to a distance of about 80 light years.

Dying red giant stars may have habitable extra-solar planets. How would ET see his or her sky from such a planet? Philip Platt, an astronomer at Sonoma State University in California, imagines the view of a planet around a red star in a globular cluster (a smaller cousin of galaxies, a globular cluster contains from tens of thousands to a million stars in a sphere about 100 light years in diameter): 'If the planet is earthlike, the sky won't be blue, because the star won't produce enough blue light. Instead, the sky might be orange or faintly red even at noon. The sky would quickly become black after sunset, allowing the glory of the cluster to emerge ... Throughout the sky would be hundreds or perhaps thousands of brilliant stars, flaming jewels ... They would cast a visible shadow and be seen easily during the daytime ... Most would be ruby red, punctuated by blue sapphires.' He acknowledges that such a planet may exist only in our imagination.

What would aliens really look like?

We have made fictional aliens in our own image, but real aliens won't look like us. Species evolve, they are not created instantly. Humans are one of the billions of species that have arisen on Earth in 3.8 billion years of evolution. There is no straight line from the origin of life to intelligent humans. There were many evolutionary branches before one pathway led to the evolution of humans. It's statistically impossible that the same maze of evolutionary pathways would be repeated if evolution on our planet was run through again.

A habitable alien planet could differ from ours in many ways – distance from its parent star, gravity, moons, chemical composition, and so on – which could force the development of radically different life forms. All this makes it absolutely impossible that one day we will say hello to a loveable, wrinkly, long-fingered ET. 'Finding another planet with our kind of dinosaurs or people is more unlikely than finding a remote Pacific island on which the natives speak perfect German or cockney rhyming slang,' says Jack Cohen, a reproductive biologist and the world's foremost authority on the physiology and behaviour of aliens.

Ulrich Walter of the Institute of Astronautics in Munich, who has made a scientific study of alien physiology, believes that the aliens are likely to have some similarities to life on our planet: they will be carbon-based life, basically composed of carbon, hydrogen, nitrogen and oxygen. As such life forms require water for their survival, their bodies need to contain water, making them soft like ours. Their bodies would also contain 'blood', a liquid to transport nutrients to all the body cells. He estimates that their size could be between 10 centimetres and 10 metres, but most likely it would be about 1 metre. They would have a skeleton to bear the body, and probably also eyes, a lung and a mouth. 'The only question that remains is how these parts are to be located,' he says. 'And this is something, the only thing, we leave to the imagination of science fiction writers and fantasy artists.'

Frank Drake believes that intelligent aliens 'won't be too much different from us ... if you saw them from a distance of a hundred yards in the twilight you might think they were human.' He says that there will be some similarities: an upright stance, a head with eyes up top for good viewing and limbs free to manipulate tools, a requisite too for a technological society. 'My fun speculation – it's serious, but it sounds like a joke – is that I think, more often than not, the species will have four arms instead of two, because we all know that two arms are not good enough when you have to carry groceries,' he says. Perhaps Drake has been influenced by the images of four-armed Hindu goddess Kali.

Cohen does not support this tongue-in-cheek view of aliens. 'The real aliens will be very alien indeed,' he says. He is almost certain that they will not be humanoid. He considers the development of spines and skeletons as 'an evolutionary accident that could well be unique to Earth'. 'We can make an educated guess that we'll never see anything like a human being on an Earth-like planet.' They will not have faces like ours at all – eyes, yes, but nose, external ears, teeth-like-scales, no – and certainly will not share our interest in prurient experiences,' he says. 'No alien pornography of our kind, then.'

There is a tendency for people to think that biochemical life is the only kind of life, he says, but he would not be surprised at all if we find other kinds of non-water-based life.

'You can find self-replicating vortexes in the atmosphere of the Sun. There are all sorts of self-reproducing systems in all sorts of environments which could hold potential for life.'

Until we meet an alien, everyone is free to design their own. You may do well if you heed Cohen's advice: 'This involves deeper, more divergent science than DNA-is-God-and-RNA-is-its-Prophet molecular biology and is much more fun, too.'

Would they have religion?

OK, ETs won't look like us, but would they have religions that resemble ours? No, says Seth Shostak; however, he is more interested in the larger question of whether the extra-terrestrials will have religion at all. 'Virtually every group of peoples on Earth has some sort of religion, and that suggests that religion is a "universal" phenomenon.' But he is worried that if we're hard-wired for religion as 'an accident of evolution', aliens might find our interest in religion a peculiar anomaly. 'This is a disturbing scenario, although not really a threatening one.'

At least Jill Tarter is not threatened by this scenario. In a thoughtful little essay, 'O give ye praise Europans', she says that organised religion is an invention of the mind, as envisioned by evolutionary biologist Steven Pinker. A techno-logical civilisation cannot live for long, she warns, unless it has 'a universal religion with no deviations, no differ-entiations – absolutely global and compelling for all'. Such a religion, she believes, might be able to co-exist for a long time with technological development without precipitating the worst of human tendencies. 'If that development is pos-sible for any civilization, then I would speculate that, if and when we ever get a message, it's going to be a missionary appeal to try to convert us all. And, on the other hand, if we get a message and it's secular in nature, I think that says that they have no organized religion – that they have outgrown it.'

Unlike Asimov's three famous laws of robotics, Arthur C. Clarke's three laws, postulated in 1962, deal with humans:

Clarke's First Law: When a distinguished but elderly scientist states that something is possible, he is almost certainly right. When he states that something is impossible, he is very probably wrong.

Clarke's Second Law: The only way of discovering the limits of the possible is to venture a little way past them into the impossible.

Clarke's Third Law: Any sufficiently advanced technology is indistinguishable from magic.

The Third Law has prompted Michael Shermer, the founding publisher of *Skeptic* magazine, to propose Shermer's Law: Any sufficiently advanced extraterrestrial intelligence is indistinguishable from God.

In an essay, 'Is God more than a sufficiently advanced extraterrestrial intelligence?' (www.edge.org), Shermer says that if we ever do find extraterrestrial intelligence (ETI) 'it will be as if a million-year-old "Homo Erectus" were dropped into the middle of Manhattan, given a computer and cell phone and instructed to communicate with us. ETI would be to us as we would be to this early hominid – godlike.'

Freeman Dyson has also expressed similar views:

'Whoever they are, they probably know a great many things that we don't know. Of course, the danger is that if they know it all, then we'd lose self-confidence.'

Although science has not even remotely destroyed religion, Shermer says, his law predicts that the relationship between the two will be profoundly affected by contact with technologically advanced extraterrestrials. 'To find out how, we must follow Clarke's Second Law, venturing courageously past the limits of the possible and into the unknown,' he suggests.

What happens when earthlings are 'abducted' by aliens?

Alien abduction stories abound in the media and on the internet. Most stories are strikingly similar and follow a common pattern. If you're ever abducted by aliens:

Encounter: You will be either in your bed or car, usually at night. If you are in your bed, you suddenly wake up and see non-human figures coming through the walls of your room or standing near your bed. You may even see a space-ship outside your window. If you are in your car, you see the car being pulled to the side by a bright object. You see strange beings exiting from a spaceship and coming towards you.

You are unable to move or speak. You see flashing lights and hear buzzing sounds.

The aliens are usually about 120 centimetres (4 feet) tall. Their bodies are thin and spindly with huge heads, slanted wrap-around black eyes, grey skin, and no hair or nose. They don't talk, but communicate with you telepathically with their large eyes.

Abduction: You are taken to the spaceship. You don't walk; you either float or are carried up by a beam of light. The inside of the spaceship looks like a high-tech doctor's examination room full of tables on which other humans lie.

Examination: You are subjected to a painful medical examination. The aliens probe you by inserting instruments into virtually every part of the body. You are subjected to various forms of unwilling sex. They extract your ovum (or sperm). Sometimes they place tiny implants in your body, especially the nose. No person has ever been found with an actual implant. Most abductees claim that they lost the nose implant when they sneezed.

Tour: The aliens may give you a tour of their spaceship or show some alien artefacts. No person has ever brought back any extraterrestrial souvenirs with them from their flying saucer tour.

Return: It's all over within a short time. You are back in your bed or car. You are confused and scared. You may find puzzling 'surgical' scars or cuts on your body and you don't remember how you got them.

UFOs: Many stories have associations with UFOs. The abductees may not have seen the UFO, but they usually read or see in the television the next day that a UFO passed over where they were when they had the experience.

Location: Most incidents happen in the United States. No one knows why the aliens always pick up Americans.

The alien-abduction phenomenon began with the case of Betty and Barney Hill. Late one night in 1961, when Betty and her husband Barney were returning to their New Hampshire home from a holiday in Canada, they noticed a bright light following their car. The light became brighter and brighter until it was clearly visible as a spaceship. Their car stalled and the spaceship landed near the car. The aliens came out and took them inside the spaceship, where they were subjected to a medical examination. Betty was given a brief tour of the spaceship and when she asked where the aliens were from she was shown a star map.

When they 'awoke', the couple continued their drive home, but without any memories of the incident. Weeks afterwards, they started experiencing nightmares and went to see a psychiatrist. Under hypnosis they 'remembered' having been abducted by aliens and subjected to painful probing of their bodies; Betty also drew the star map shown by the aliens. A few years later, a UFO researcher claimed that Betty's vague star map resembled Zeta Reticuli, a binary star system located in the constellation Reticulum, 39 light years away. Betty and Barney's story became highly popular when John Fuller told it in his book, *The Interrupted Journey* (1966). It was later made into the television movie *The UFO Incident* (1975), in which Barney was played by James Earl Jones. The book and the movie started a new genre, and the publishing and entertainment industries continue to milk this cash cow.

At least one eminent scientist took these strange happenings seriously. John Mack, a Pulitzer Prize-winning psychiatrist at Harvard University who died in 2004, was ridiculed by the scientific community when he claimed in his book, *Abduction: Human Encounters with Aliens* (1994), that there was no psychological explanation for the phe-nomenon. He argued that abductees – he worked with 200 of them and called them 'experiencers' – were not crazy and their experiences were genuine. 'It's both literally, physically happening to a degree; and it's also some kind of psychological, spiritual experience occurring and originating perhaps in another dimension ... I would never say, yes, there are aliens taking people. But I would say there is a compelling powerful phenomenon here that I can't account for in any other way, that's mysterious. Yet I can't know what it is but it seems to me that it invites a deeper, further inquiry.'

Susan Clancy, a Harvard University psychologist and the author of *Abducted: How People Come to Believe They Were Kidnapped by Aliens* (2005), agrees that these people are not crazy, but they do have a tendency to fantasise and to hold unusual beliefs and ideas. 'They do not only believe in alien abduction, but also things like UFOs, ESP, astrology, tarot, channelling, auras and crystal therapy,' she says. 'They also have in common a rash of disturbing experiences for which they are seeking an explanation. For them, alien abduction is the best fit.' She ties the phenomenon to sleep paralysis, a condition in which the usual separation between sleep and wakefulness gets out of synchronisation. Sleep paralysis occurs when the body is in the dream phase of sleep and it disconnects from the brain.

There are other psychological explanations as well. Some psychologists believe that alien abductions and other mystical and psychic experiences may be linked to excessive bursts of electrical activity in the temporal lobes. These lobes – one on each side of the brain, located near the ears – control hearing, speech and memory. Some argue that alien abduction may be disguised memories of sexual abuse.

'Many of us long for contact with the divine, and aliens are a way of coming to terms with the conflict between science and religion,' says Clancy. 'I agree with Jung: extra-terrestrials are technological angels.'

Is an alien visit possible technically?

How long will it take these 'technological angels' to travel to Earth from their home, say, in Zeta Reticuli? Thirty-nine years if they travel at the speed of light. Einstein has banned travel at this speed, at least for earthlings. His special theory of relativity says that the mass of a moving object increases as its speed increases. At the speed of light, which is about 300,000 kilometres per second, the mass becomes infinite, and therefore nothing can move faster than light.

But this Einsteinian stricture hasn't stopped scientists speculating about faster-than-light travel through 'worm-holes' in space-time. Interestingly, the idea of wormholes was made possible by Einstein's theory of general relativity in 1916. However, scientists didn't take much interest in them until Carl Sagan employed a wormhole in *Contact* to transport Dr Ellie Arroway to the centre of the galaxy and back.

Our universe appears as space-time: three dimensions in space (up-down, left-right, and forward-backward) and a fourth dimension known as time. Wormholes are 'tunnels' between two different places in space and time. Imagine a worm crawling on a flat leaf from point A near one edge to point B near the other edge. If the leaf is curled up, the worm will take less space (and time) to crawl through the 'shortcut tunnel' between points A and B. The idea seems simple in theory, but the wormholes scientists talk about are extremely unstable, short-lived and narrow (not even wide enough for a proton to pass through). There is another small matter: the nearest wormhole gateway is through the massive black hole at the centre of the galaxy. If 'technological

WHY AREN'T THEY HERE?

angels' – which may be thousands, millions or billions of years ahead of us – have found ways of breaking our laws of physics, a new Wormhole Tunnel Express station may soon open near your bed or car. You have been warned.

As far as we earthlings are concerned, wormholes are like the rabbit-hole in *Alice in Wonderland*. However, Michio Kaku believes that faster-than-light travel is probably attainable in 100,000 to 1 million years. 'When I look at the age of the universe, I see that we've attained technology in the blink of an eye. There's plenty of time.'

Some scientists are impatient. These scientists – for whom 'wildly optimistic dreamers' is virtually a job description, according to the *New York Times* – are not willing to wait for a million years to colonise other worlds. They pin their hopes on what they call multi-generational space travel in a rocketless spaceship. The spaceship will be equipped with ultra-thin, ultra-large sails which will be propelled by a beam of laser light from Earth or high-energy particles from the sun-soaked planet Mercury. The spaceship will travel at about 30 per cent of the speed of light, or about 88,000 kilometres per second. Compare it with the speed of Voyager, the fastest man-made object in space, which is about 15 kilometres per second. The spaceship would take about 50 years to reach the nearest star, Alpha Centauri, about 4.4 light years away. Who would want to spend their next 50 years in a spaceship to reach Alpha Centauri to meet alien rocks, or even alien rock stars?

One of these 'wildly optimistic dreamers' is John Moore, an anthropologist at the University of Florida. He dreams of ships powered by multi-generational crews – people who man the ship with their children, grandchildren and descendants through six, eight or ten generations. Moore's computer simulations show that a founding crew of 80 to 100 people would provide enough genetic variability to make it viable for more than a thousand years.

If intelligent extraterrestrials would have thought of this *Star Trek* plan, they would be here; and Fermi would not have asked his question.

Rasmus Bjørk of the Niels Bohr Institute in Copenhagen has taken a down-to-earth approach to 'investigate the possible use of space probes to explore the Milky Way, as a means to both finding life elsewhere in Galaxy and as finding an answer to the Fermi paradox'. He says that the use of interstellar probes is the only way to perform detailed investigations of extra-solar planets believed to harbour life. His calculations show that eight probes – travelling at a tenth of the speed of light and each capable of launching eight sub-probes – would take 100,000 years to explore 40,000 star systems in the habitable zone of the galaxy. He agrees that exploring the galaxy by space probes is 'horribly slow'.

Intelligent extraterrestrials seem to agree with Bjørk; and that's why Fermi asked the question:

Why aren't they here? ...

Some scientists believe that microbes may have travelled from Mars to Earth, and also in the opposite direction. Alien microbes are probably already here, and therefore we do not have to find any excuses for their absence. But we have made numerous excuses on behalf of intelligent extraterrestrials as to why they are not here. Here are some more excuses:

They have visited, but not in recorded history.

They have visited in recorded history (the Tunguska event of 1908? – see the author's *The Mystery of the Tunguska Fireball*; Erich von Däniken's ancient astronauts? – see *Chariots of the Gods*).

They choose to exist in other dimensions.

They are intelligent but lack the insatiable curiosity that humans display. They have no desire to communicate.

They are omniscient, omnipotent beings and do not want to come here, as they fear we might start worshipping them like gods.

They find travelling by spaceships boring; they prefer tele-conferencing by telepathy.

They are couch potatoes with sophisticated remote controls (why leave your couch, let alone your planet?).

'They came, they had a look around and decided that it wasn't for them after all – and that we're altogether too ghastly to warrant a waste of another nanosecond of space-time. And so they have gone back wherever it was they came from, annoyed, depressed and possibly suicidal.' This is the dispiriting view of *The Spectator* magazine's Rod Liddle.

... they are already here and they came by UFOs ...

Most UFO fans believe that we've already discovered extra-terrestrial intelligent beings and they are visiting Earth. They also believe that governments are covering up this fact because they know it would trigger a panic; governments are afraid of admitting something that is beyond their control. American, Canadian and British opinion polls during the past decade have also supported this view. The polls have shown that on average, 50 per cent of the public believe we have been visited by aliens.

Most of these UFO fans have also claimed to have seen a UFO. What have they seen? Planes, jets, helicopters, balloons, strange flocks of birds. Unusual light patterns caused by astronomical and meteorological phenomena. Optical illusions caused by smoke and dust. Psychological delusions. Deliberate hoaxes.

Many people who have had vivid memories of close encounters with UFOs are not convinced by these explanations. A report by the British Ministry of Defence on 'Unidentified Aerial Phenomena' (the report was completed in 2000 and made public for the first time in 2006) says that some 'events are almost attributable to physical, electrical and magnetic phenomena in the atmosphere, mesosphere and ionosphere'. These events create regions of electrically charged plasma which appear as bright, fast-moving objects to observers. They are probably caused by a meteorite entering the atmosphere, 'neither burning up completely nor impacting as meteorites, but forming buoyant plasmas'. Sometimes the field between certain charged buoyant objects forms an area, often triangular, which does not reflect light. This explains why some UFOs are described as black spaceships, often triangular, and up to hundreds of metres in length. The events cannot be detected by radar.

The 400-page report also claims that 'the close proximity of plasma related fields can adversely affect a vehicle or person'. It has been medically proven that local electromagnetic fields can cause responses in the temporal lobes of the human brain. 'These result in the observer sustaining (and later describing and retaining) his or her own vivid, but mainly incorrect, description of what is experienced,' the report says. 'This is suggested to be a key factor in influencing the more extreme reports found in the media and are clearly believed by the "victims".'

Such events are so infrequent that they are unique to most observers. In that sense they are truly encounters of the UFO kind, but they are not encounters with extra-terrestrials. 'U' in UFO simply means 'unidentified'; it doesn't suggest 'extraterrestrial'.

Carl Sagan gives another explanation: 'If each of a million advanced technical civilizations in our galaxy launched at random an interstellar spacecraft each year, our solar system would, on average, be visited once every 100,000 years.'

He employs his 'Santa Claus hypothesis' to explain UFOs. The hypothesis maintains that in a period of eight hours or so on 24–25 December of each year, an elf visits 100 million homes in the United States. If this elf spends one second per house, he has to spend three years just filling the stockings in all the houses. Even with relativistic reindeer, the time spent in 100 million houses is three years and not eight hours. 'We would then suggest that the hypothesis is untenable,' he says. 'We can make a similar examination, but with greater certainty, of the extraterrestrial hypothesis that holds that a wide range of UFOs viewed on the planet Earth are space vehicles from planets of other stars.'

... and by rain

If intelligent aliens are visiting Earth by UFOs, microbial aliens are travelling by rain to visit Earth, at least a tiny part of it in the state of Kerala in India. On the morning of 25

July 2001, residents of a small town in the Kottayam district heard a loud sonic boom accompanied by lightning. What followed was a three-hour spell of heavy monsoon rain. For about fifteen minutes the rain turned blood red, which eye-witnesses claim was falling as scarlet sheets. Many people in the street found their clothes turned pink. Some even noticed that the rain burned leaves on trees. Similar blood-red rain bursts during normal rains continued sporadically there and in other areas along the coast for about two months.

Explanations provided by local experts included: a burning meteorite threw out red dust which came down as red rain; fine dust blown from Arabian deserts got mixed with monsoon rains; red rain particles were possibly fungal spores from trees; and a fine mist of blood cells produced by a meteor striking a high-flying flock of bats.

The red rain would have been forgotten completely if Godfrey Louis, a solid-state physicist at Mahatma Gandhi University in Kottayam, had not decided to investigate this mysterious phenomenon with his student Santhosh Kumar. From various news reports and other sources, the research-ers compiled a list of 124 incidences of red rain, and collected samples of red rainwater from various places more than 100 kilometres apart.

Their analysis of the rainwater shows that the particles are about 4 to 10 micrometres in size, almost transparent red in colour, and well dispersed in the water. The particles have an appearance similar to single-celled organisms. One chemical analysis has shown that they have about 50 per cent carbon, 45 per cent oxygen and traces of sodium and iron. All this data is consistent with biological materials. But the cells lack any nucleus and DNA. Life as we know it must contain DNA. Are they alien microbes?

Louis and Kumar conclude that the sonic boom heard by many people before the rain was triggered by a meteorite. When it disintegrated in the upper atmosphere, it shed the embedded alien microbes, which were mixed with the clouds and then fell as rain.

Chandra Wickramasinghe, a champion of the panspermia theory (see page 57), and his colleagues at Cardiff University have examined the particles and have confirmed that they are biological cells. Their investigation also suggests the presence of DNA. 'However, this identification is not yet fully confirmed,' they say. Milton Wainwright, a microbiologist at the University of Sheffield, who has also analysed the samples, claims that the cells are 'morphologically similar to fungal spores or algae'. He rules out dust, sand, fat globules or blood.

Terrestrial or alien? Even if terrestrial, how did it happen? Scientists are still debating, but the 'alien rain' has already found a home on thousands of alternative science websites and blogs. *X-Files* fans don't like to wait for scientific verdicts.

EPILOGUE

'They Could Show Up Here Tomorrow'

Why are humans fascinated by life beyond Earth? Aristotle answered this question more than two millennia ago when he said: 'All men by nature desire knowledge.' In our millennium, Paul Davies has said: 'For in a sense, the search for extraterrestrial life is really a search for ourselves – who we are and what our place is in the grand sweep of the cosmos.'

Our place in the grand sweep of the cosmos will never be the same, even if we discover some fungus on another planet. It will show that probably life is widespread throughout the universe. The ages-old hubris that our planet is special will disappear. Not completely. Some will claim that humans are still the only intelligent life.

William Whewell, an early voice in the extraterrestrial life debate, said in 1853: 'The discussions in which we are engaged belong to the very boundary regions of science, to the frontier where knowledge ... ends and ignorance begins.' His words still echo reality after more than a century and a half, when we talk about technology-savvy aliens that have nothing else to do but flash their business cards by radio waves (or some such means we're familiar with). The search for extraterrestrial intelligence is based on this premise.

Though science has made great strides in finding primitive life, or at least likely abodes of such life, our searches are still limited by the terrestrial concept of life and intelligence. 'We are good at looking for things like ourselves,' comments American science writer Dennis Overbye.

Dreamers employed in the movie and advertising industries aren't waiting for the momentous discovery of aliens. They have already signed up aliens of all kinds, cute or monster. Mark Dolliver of *Adweek* magazine describes a television commercial for a New York fitness club that shows a paunchy man sitting on a bench in Central Park eating his lunch. 'Suddenly, a flying saucer hovers above him, and its pilots – two curvaceous space gals clad in outer-space Gaultier – try to beam him aboard ... Sadly, his weight is more than the saucer's beam can hoist. So, when a trim athletic man jogs into view, the women shift the beam his way, and he's last seen rising into the spaceship with a big grin on his face.' He asks: 'If that prospect doesn't motivate a person to get into shape, what on earth would?'

SETI researchers keep on reminding us that 'absence of evidence is not evidence of absence'. Perhaps that's why Frank Drake, the pioneer researcher, is optimistic that 'they could show up here tomorrow'. When they do show up, you know that trim and athletic natives have a better chance of being picked up. Where's that exercise bike?

And if you're lucky enough to be picked up, you might like to ask them some questions. In its Christmas 1996 issue, *New Scientist* magazine published the results of a competition: 'What ten questions would you ask an alien if you met one?' A selection from the winning entries, with the name of the contributor in brackets:

'Do you really want to be taken to our leader, or would you rather go to the pub?' (John Rowland)

'So you got rid of all the bacteria on your world by sending them to Earth on a meteorite, did you?' (Karen Pike)

'Have you ever been abducted by aliens?' (Gordon Stewart)

What are your questions?

Selected Bibliography

This list does not include sources that have been clearly identified within the text.

Luc Arnold, 'Transit light-curve signatures of artificial objects', *The Astrophysical Journal*, 1 July 2005

Svante Arrhenius, *Worlds in the Making: The Evolution of the Universe*, trans. H. Borns, Harper & Brothers, New York, 1908

John A. Ball, 'The zoo hypothesis', *Icarus*, vol. 19, pp. 347–9, 1973

Rasmus Bjørk, 'Exploring the Galaxy using space probes', arXiv:astro-ph/0701238 v1, 9 January 2007 (http://arxiv.org)

Susan Blackmore, 'Alien abduction: inside story', *New Scientist*, 19 November 1994

Ronald N. Bracewell, *Galactic Club: Intelligent Life in Outer Space*, W.H. Freeman, San Francisco, 1974

Milič Čapek, *The Philosophical Impact of Contemporary Physics*, Van Nostrand, Princeton, 1961

Max Caspar, *Kepler*, trans. and ed. C. Doris Hellman, Abelard-Schuman, London, 1959

John L. Casti, *Paradigms Lost: Images of Man in the Mirror of Science*, William Morrow, New York, 1989

Marcus Chown, 'Is anybody out there?', *New Scientist*, 23 November 1996

Gale E. Christianson, 'Kepler's Somnium: science fiction and the renaissance scientist', *Science Fiction Studies*, vol. 3, part 1, no. 8, March 1976

Paul Clancy, André Brack and Gerda Horneck, *Looking for Life, Searching the Solar System*, Cambridge University Press, Cambridge, 2005

Giuseppe Cocconi and Philip Morrison, 'Searching for interstellar communication', *Nature*, 19 September 1959

Jack Cohen, 'How to design an alien', *New Scientist*, 21–28 December 1991

Jack Cohen and Ian Stewart, 'Where are the dolphins?', *Nature*, 22 February 2001

Francis Crick and Leslie Orgel, 'Directed panspermia', *Icarus*, vol. 19, pp. 341–6, 1973

Paul Davies, 'The harmony of the spheres', *Time*, 19 February 1996

—— 'Do we have to spell it out?', *New Scientist*, 7 August 2004

Bernard de Fontenelle, *Conversation on the Plurality of Worlds*, trans. H. A. Hargreaves, University of California Press, Berkeley, 1990

Steven J. Dick, *Life on Other Worlds: The 20th-Century Extraterrestrial Debate*, Cambridge University Press, Cambridge, 1998

—— 'They aren't who you think', *Mercury*, November–December 2003

Mark Dolliver, 'Our better natures, happy brides in tears, extraterrestrials etc.', *Adweek*, 10 November 1997

Frank Drake, 'The E.T. equation, recalculated', *Wired*, December 2004

Freeman Dyson, 'Gravity is cool, or, why our universe is hospitable to life', Oppenheimer Lecture, given at the

University of California in Berkeley, 9 March 2000 (http://www.hartford-hwp.com/archives/20/035.html)

Jerry E. Ehman, 'The Big Ear Wow! Signal: What we know and don't know about it after 20 years', 3 February 1998 (http://www.bigear.org/www.20th.htm)

Murray Gell-Mann, 'Nature conformable to herself', *Complexity*, vol. 1, no. 4, 1995

Louis Godfrey and Santhosh Kumar, 'The red rain phenomenon of Kerala and its possible extraterrestrial origin', arXiv:astro-ph/0601022 v1, 2 January 2006 (http://arxiv.org)

J. Richard Gott, 'Implications of the Copernican principle for our future prospects', *Nature*, 27 January 1993

Gary Hamilton, 'Mother Superior', *New Scientist*, 13 September 2005

Michael J. Hart, 'An explanation for the absence of extra-terrestrials on Earth', *Quarterly Journal of the Royal Astronomical Society*, vol. 16, pp. 128–35, 1975

Roger Highfield, 'The greatest discovery of all time', *Daily Telegraph*, UK, 5 October 2005

Fred Hoyle and N. Chandra Wickramasinghe, *Life on Mars: The Case for a Cosmic Heritage*, Clinical Press, Bristol, 1997

S. Hsu and A. Zee, 'Message in the sky', arXiv:physics/0510102 v2, 6 December 2005 (http://arxiv.org)

James Jeans, 'Is there life on other worlds?', *Science*, vol. 95, p. 589, 1942

Michio Kaku, 'The Physics of Extra-Terrestrial Civilizations', 2003 (http://www.mkaku.org)

N.S. Kardashev, 'Transmission of information by extra-terrestrial civilizations', *Soviet Astronomy*, vol. 8, no. 217, 1964

Günter von Kiedrowski, 'Primordial soup or crêpes?', *Nature*, 2 May 1996

John Lear, *Kepler's Dream, with the full text and notes of* Somnium, sive Astronomia lunaris, *Joannis Kepleri, trans. Patricia F. Kirkwood*, University of California Press, Berkeley, 1965

Charles H. Lineweaver and Tamara M. Davis, 'Does the rapid appearance of life on Earth suggest that life is common in the universe?', arXiv:astro-ph/0205014 v2, 8 May 2003 (http://arxiv.org)

Ralph Lorenz, 'Death of a watery world', *New Scientist*, 20 September 1997

Lucretius, *De rerum natura*, trans. R.C. Trevelyan, Cambridge University Press, Cambridge, 1937

Francis Lyall, 'SETI and the law: what if the search succeeds?' *Space Policy*, vol. 14, pp. 75–7, 1998

Ernst Mayr, 'The probability of extraterrestrial intelligent life', in E. Regis (ed.), *Extraterrestrials: Science and Alien Intelligence*, Cambridge University Press, Cambridge, 1985

Hazel Muir, 'It's raining aliens', *New Scientist*, 4 March 2006

J. Madeleine Nash, 'How did life begin?', *Time*, 11 October 1993

Dennis Overbye, 'Where are those aliens?', *New York Times*, 11 November 2003

Paul Parsons, 'Dusting off panspermia', *Nature*, 19 September 1996

Clifford Pickover, *The Science of Aliens*, Basic Books, New York, 1998

Philip Plait, 'Under alien skies', *Astronomy*, January 2003

Peter Radetsky, 'Life's crucible', *Earth*, February 1998

Christopher Rose and Gregory Wright, 'Inscribed matter as an energy-efficient means of communication with an extraterrestrial civilization', *Nature*, 2 September 2004

Carl Sagan, *The Cosmic Connection: An Extraterrestrial Perspective*, Doubleday, New York and Cambridge, 1973

Carl Sagan and Frank Drake, 'The search for extraterrestrial intelligence', *Scientific American*, May 1995

Carl Sagan, Linda Salzman Sagan and Frank Drake, 'A message from earth', *Science*, vol. 175, p. 881, 1972

Sundar Sarukkai, *Translating the World: Science and Language*, University Press of America, Lanham, 2002

Jeff Secker, Paul Wesson and James Lepock, 'Astrophysical and biological constraints on radiopanspermia', arXiv: astro-ph/9607139 v1, 26 July 1996 (http://arxiv.org)

William Sheehan, Nicholas Kollerstrom and Craig B. Waff, 'The case of the pilfered planet: did the British steal Neptune?', *Scientific American*, December 2004

I.S. Shklovskii and Carl Sagan, *Intelligent Life in the Universe*, Helden-Day, San Francisco, 1966

Seth Shostak, 'The day after', *Mercury*, March–April 1996

——— 'Religion, science and ET', *Mercury*, May–June 1998

——— 'The Future of SETI', *Sky & Telescope*, April 2001

——— 'Drake's brave guess', *Discover*, May 2006

Walter Simmons, Sandip Pakvasa, John Learned and Xerex Tata, 'Timing data communication with neutrinos – a new approach to SETI', *Quarterly Journal of the Royal Astronomical Society*, vol. 35, p. 321, 1994

Walter Simmons and Sandip Pakvasa, 'Sending signals while remaining hidden', *Mercury*, May–June 2003

Richard E. Smalley, 'Of chemistry, love and nanobots', *Scientific American*, September 2001

The Staff at the National Astronomy and Ionosphere Center, 'The Arecibo message of November 1974', *Icarus*, vol. 26, p. 463, 1975

W.T. Sullivan (ed.), *The Early Years of Radio Astronomy: Reflections Fifty Years after Jansky's Discovery*, Cambridge University Press, Cambridge, 1984

George W. Swenson, Jr., 'Intergalactically speaking', *Scientific American*, July 2000

Jill Tarter, 'O give ye praise Europeans', *Free Inquiry Magazine*, vol. 20, no. 3, Summer 2000

Jill Tarter and B. Zuckerman, 'Is anyone out there?', *Nature*, 18 October 1979

Frank. J. Tippler, 'We are alone in our galaxy', *New Scientist*, 7 October 1982

Neil deGrasse Tyson, 'Exoplanet Earth', *Natural History*, February 2006

Peter Ulmschneider, *Intelligent Life in the Universe: From Common Origins to the Future of Humanity*, Springer, Berlin, 2003

Elling Ulvestad, 'Biosemiotic knowledge – a prerequisite for valid exploration of extraterrestrial intelligent life', *Sign System Studies*, vol. 30, no. 1, 2002

Douglas Vakoch, 'Altruism is the key to interstellar communication', *Research News and Opportunities in Science and Theology*, vol. 2, no. 3, November 2001

Surendra Verma, *The Little Book of Scientific Principles, Theories & Things*, New Holland Publishers, Sydney, 2005

Ulrich Walter, 'Sind wir allein im all?', *StarObserver Spezial*, March 1999

Peter D. Ward and Donald Brownlee, *Rare Earth: Why Complex Life Is Uncommon in the Universe*, Copernicus, New York, 2000

David Warmflash and Benjamin Weiss, 'Did life come from another world?', *Scientific American*, November 2005

Stephen Webb, *Where Is Everybody? Fifty Solutions to Fermi's Paradox*, Copernicus, New York, 2003

N. Chandra Wickramasinghe and Fred Hoyle, 'Evolution of life: a cosmic perspective', ActionBioscience.org, May 2001 (http://www.actionbioscience.org)

T. L. Wilson, 'The search for extraterrestrial intelligence',
  Nature, 22 February 2001
'Why no SETI signals yet?', *Sky & Telescope*, November 2004

Acknowledgements

I'm grateful to Dr Ray Kurzweil (www.KurzweilAI.net) and Professor Sundar Sarukkai (National Institute of Advanced Studies, Bangalore) for commenting on their research, and to Drs Yvan Dutil and Stéphane Dumas (Defence Research Establishment Valcartier, Canada), Dr Jerry ('Wow!') Ehman (North American Astrophysical Observatory) and Ngan Truong (SETI Institute, California) for providing illustrations.

My special thanks go to my mate Eric ('Fiz') Fiesley and the mob (*They're a Weird Mob*) on the Australian Skeptics' Qskeptics online forum, especially Steve Roberts, Fred Thornett and Barry Williams, for discussing maths and Martians on Qskeptics; my colleague Simon Kwok for his help with illustrations; and to my old friend Jitendra Sharma, a well-known Hindi writer, for his continuing encouragement.

Once again, I'm indebted to Icon Books' publishing director Simon Flynn for giving me the opportunity to write this book, and their editorial director Duncan Heath for his down-to-earth advice.

Finally, I would like to thank my wife Suman and my sons Rohit and Anuraag for their unfailing support during the days I was an alien to them.

Australian Government | **Australia Council** for the Arts

This project has been assisted by the Australian Government through the Australia Council, its arts funding and advisory body.

Index

Page numbers in italics indicate illustrations